THE SPY
NEXT DOOR

THE SPY
NEXT DOOR

THE EXTRAORDINARY SECRET LIFE OF ROBERT PHILIP HANSSEN, THE MOST DAMAGING FBI AGENT IN U.S. HISTORY

Elaine Shannon
and Ann Blackman

LITTLE, BROWN AND COMPANY
Boston   New York   London

First Edition

For information on AOL Time Warner Book Group's online publishing program,
visit www.ipublish.com.

ISBN 0-316-71821-1
LCCN 2001077558
10 9 8 7 6 5 4 3 2 1
Q-Mart

Printed in the United States of America

To Edward Hogan Shannon and Kathryne Wallace Shannon,
who showed us the God of selfless love, simple justice,
plain truth, and cosmic jokes

and

To Mike, Leila, and Christof,
who hear the song of the loon on the Damariscotta

CONTENTS

THE SPY
NEXT DOOR

PROLOGUE

Robert Hanssen was the quintessential suburban dad, working a government job to pay for a four-bedroom, split-level house on a cul-de-sac in a modest neighborhood of suburban Virginia. Seemingly devoted to his wife, Bonnie, Hanssen sent all six of his kids to Catholic schools and Catholic colleges. Three aging cars sat in his driveway, and an electric fence ran across the front lawn to keep the family's black lion of a dog from roaming. Unlike Bonnie, who was popular and funny, Hanssen was not particularly friendly, but the Virginia suburbs were full of couples made up of buoyant wives and boring husbands. Neighbors often saw Hanssen walking alone through the park behind his house at night, taking his dog out for a romp. A few may have wondered how he made ends meet on a government salary that ranged from $87,000 to $114,000, but these were the American suburbs — who knew what anyone did? Nobody would have guessed that this dad with a clenched smile had a very secret life. Hanssen may have professed a repeated desire to walk in the path of Jesus, but his biblical model was, in fact, someone else.

Robert Hanssen was an American Judas, one of the most damaging spies ever to work against the United States.

What is it that makes someone run from all that is comfortable and familiar? What is it that snaps in the human spirit, causing a person to break the bonds of family and friends, faith and country, betraying all and everyone that ever mattered? What allows a man who professes to be deeply religious to spy for a political system that sought to repress religious expression? How does a man who leaves the office early to attend anti-abortion rallies allow himself to snitch on fellow spies, knowing they might well be executed? And on and on, a thicket of paradoxes and questions that call up still other questions.

During the period that Robert Hanssen betrayed his country, scores of employees of the federal government and its contractors were convicted of espionage. It would take years to measure the scope of Hanssen's treachery, but the early assessments were that only CIA turncoat Aldrich Ames and Navy code specialist John Walker had done as much damage. While the Cold War was long over, there was no guarantee that the secrets Hanssen had purloined stayed in Moscow and did not fall into the hands of the intelligence services of the People's Republic of China, North Korea, Iran, Iraq, Syria, Libya, or Cuba. "The Russians are in a position to sell, use, or trade that information to help others neutralize U.S. intelligence," said Roy Godson, director of the National Strategy Information Center, a Washington think tank that specializes in security matters. "The Russians have taught or traded their knowledge of us to other countries such as Iraq, or to terrorist and revolutionary groups." It seemed extremely unlikely that even the most disgruntled ex-KGB officer would deal with the Taliban regime in Afghanistan, where the Soviet army was shredded over a bloody decade. But on September 11, 2001, when a team of Osama bin Laden followers staged spectacularly successful attacks on New York and Washington, forever altering the landscape of American life, who dared say that anything was impossible?

Until the cataclysmic events of September 11, the day Robert Hanssen was arrested was the FBI's darkest moment in a decade during which the FBI's fiascoes — among them, Ruby Ridge, Waco, Richard Jewell, Wen Ho Lee, and the delay of the execution of Oklahoma City bomber Tim McVeigh — were better known than its successes. For twenty-one years, through the terms of four presidents and three FBI directors, Bob Hanssen had everyone fooled. He had sold out right under the noses of the men and women who were supposed to be keeping watch.

Those who knew and worked with Hanssen say he was a bright but brittle man, that he never measured up to his father's dreams for him, and that even in childhood he was always considered a little bit different. As an adult, he kept a distance from his peers. Many FBI employees who knew Hanssen over the years and watched him teeter on rungs of the ladder they had handily climbed suspect he felt unappreciated, abused, overlooked. As younger men and women leapfrogged over him to more prestigious jobs, they say he may have become increasingly alienated, starved for the respect and attention he was convinced he deserved. They believe Hanssen saw himself as a master gamesman, a guy labeled weird and geeky who got his kicks by outsmarting those around him, a double agent with a double life who eventually got lost in the doubts and shadows of a dangerous, make-believe world. But this is all hindsight.

Hanssen held Top Secret clearance — and higher — for more than two decades without ever undergoing a polygraph or in-service psychological exam. While periodic lie-detector tests were routine at the CIA and other intelligence agencies, the FBI's top managers, including then–FBI director Louis Freeh, constantly rebuffed the pleas of lower-level chiefs to require those assigned to work with classified material to take polygraph exams. The FBI's leaders clung to the belief that the tight fabric of the organization's fraternal bonds would shield its members from temptation. In the rare case where a person of weak character slipped into the ranks, they trusted his supervisors and peers to sniff him out before he did much harm. Surely, they thought, highly trained professional investigators with keenly honed instincts could do better than some electronic gizmo.

They were wrong.

If, like a gunman on a murder-suicide mission, Hanssen signaled his distress, no one noticed. If he broadcast his intent, no one read the signs — not his family, not his friends, not his church, not his physicians, not his colleagues, not the FBI. Hanssen may simply have looked around and whispered to no one in particular, "I'll show you."

1

DANCING WITH FIRE

Bob Hanssen started the morning of February 18, 2001, much like any other Sunday. His wife, Bonnie, had fixed scrambled eggs to serve with glazed Dunkin' Donuts, Bob's favorite. With the two youngest of their six children, the family had piled into their aging beige Volkswagen van in time to join other worshipers for the 10:30 mass at St. Catherine of Siena Catholic Church in Great Falls, Virginia. The Hanssens were serious about their faith. Instead of attending services at their local parish, they made the weekly eight-mile trip because St. Catherine's was the only church in the Arlington diocese that still held a Latin mass. Their son Mark had been an altar boy there for several years.

Arriving just in time, the Hanssens took their usual seats in the third-row pew, directly in front of the large wooden statue of the Blessed Mother, *Our Lady of Sorrows*. Before sitting down, each genuflected, then knelt in prayer. Supreme Court Justice Antonin Scalia, father of nine, was sitting a few rows back and to the left. FBI Director Louis Freeh, father of six, also a member of the parish, had been to an earlier mass.

As the thunderous sound of the Rodgers electric organ signaled the start of the Mass, voices of the choir rose in unison singing "Lord of All Nations, Give Me Grace." The Reverend Franklyn Martin McAfee followed the processional to the altar, the creases of his ornate green chasuble glittering in the reflected light of candles being carried by the altar boys. Shaking a silver canister back and forth, McAfee waved nonallergenic incense (it was lily-of-the-valley that week, other times rose or geranium) over the altar. Then, turning to address the parishioners, he chanted in his familiar baritone: "*In nomine patris et filii et spiritus sancti.*" In the name of the Father, Son, and Holy Spirit. His listeners joined in the ritual by responding: "*Dominus vobiscum.*" The Lord be with you.

Most of those who attend St. Catherine's live in the Virginia suburbs, just outside of Washington, D.C. Some church members are very affluent, arriving for mass by limousine. The church population is largely white and traditional. Many have Ivy League educations and are influential in government and industry. On an average Sunday, the St. Catherine's church collection plate brings in about $15,000, and donations to the poor box, which are distributed to various charities, add a bit more. Church members are requested to put their weekly donation in envelopes marked with each family's name and address so the church can keep track of its roster. It is church policy that members from outside the parish, like the Hanssens, always list their particulars, and they are frequently reminded about it. Yet the Hanssens rarely used the envelopes. Church officials said that the family contributed some money to the parish, but not much. And when he did offer something, Bob Hanssen seemed to prefer cash with no indication as to the donor. Since the highest form of charity may be that of the anonymous donor, Hanssen may have considered this a high-minded gesture.

At the end of the ninety-minute mass on this Sunday, Father McAfee rose from his chair, adjusted his black biretta on his thick white mane, and led his church members in prayer: "O prince of the heavenly hosts, by the power of God, cast into hell Satan and all the evil spirits who wander about the world seeking the ruin of souls. Amen."

"Amen," sang several hundred parishioners in unison.

When Bob Hanssen arrived home at midday, he spent a few minutes on the lawn tossing a Frisbee to the family dog, a friendly black Labrador retriever mix named Sundae. Then he fell into conversation with a weekend guest, Jack Hoschouer, his best friend of more than forty years. At fifty-seven, Jack was as fit and energetic as a Tom Clancy hero, a retired lieutenant colonel in the Army, living in Germany, where he worked as a sales representative for an ammunition company. The two men were as close as brothers, the kind of old friends who could still laugh over those embarrassing little episodes of childhood that most people try to forget. Bob handed Jack a dog-eared copy of a book called *The Man Who Was Thursday* by G. K. Chesterton, the British journalist and conservative social philosopher best known for his detective stories. Hanssen, an avid reader with a weakness for spy novels, told his friend that this was a favorite tome, and it certainly looked much-read. The cover was torn off, the pages brittle with age and long since turned the shade of steeped tea. Written in 1908, the book tells the story of an undercover spy ring composed of seven men, all Catholic, each one code-named for a different day of the week. They live in a suburb of London called Saffron Park, a place that was "not only pleasant but perfect," call their society the "General Council of the Anarchists of Europe," and pride themselves on being enemies of society. Their mission is to kill the president of France. The gentleman who holds the position of Thursday thinks of himself as a poet and is, in his other life, an undercover policeman.

Hoschouer, who had a plane to catch, didn't give the book much thought before tossing it on top of his suitcase.

Shortly before four P.M., Bob told Bonnie that he would drop Jack at Dulles International Airport. Hoschouer had a six-o'clock flight to Phoenix to see his mother, and the airport was about a fifteen-minute drive from the house. The two men got into Hanssen's four-year-old silver Taurus. It was a routine they had followed many times before. But when they arrived at the departure terminal, Hoschouer was surprised that Bob did not offer to join him inside for a Coke, which was their usual practice. Then Jack remembered that the Hanssens' eldest daughter, Jane, was expected with her four children for Sunday supper and figured Bob simply had to get home. Jack knew also that Bob wanted to have enough time that evening to watch the Daytona 500. The two friends had been race car fans since high school and were rooting for seven-time NASCAR Winston Cup–winner Dale Earnhardt. A quick good-bye was in order. "It was an uneventful and unremarkable day," Hoschouer recalled.

What Jack did not know was that the FBI had followed Hanssen to the airport. Had Hanssen boarded an international flight, it would not have taken off. Had he boarded a domestic flight, agents planned to fly with him. But Hanssen did neither. Instead, he headed home.

But a few blocks before he got there, he made a slight detour.

The afternoon was cold and somewhat dreary, the clouds gathering in the winter sky like gray goose feathers, threatening rain. At exactly 4:34 P.M., Hanssen climbed out of his car, closed the door, and walked quickly across Crossing Creek Road toward the wooded entrance of Foxstone Park, a narrow strip of tulip poplar trees separating one suburban housing development from another. Reaching into his pocket, he pulled out a roll of white adhesive tape and placed a small vertical strip on the dark signpost. Both sides of the long, asphalt path were lined with deep puddles, pungent with the smell of moldy leaves. The sun was falling, and he had no time to waste.

Bob Hanssen was a tall man with big shoulders and a thick neck, his dark hair only slightly flecked with gray. He was somewhat stooped, his head jutting so far forward that he almost resembled a question mark. On this day, his stride was quick and purposeful. He was five weeks short of retiring from the FBI and had been angling for a job with a Russian defector named Viktor Sheymov, who had started a small security software company called Invicta Networks, located in Herndon, Virginia. Sheymov had turned him down, but Hanssen was mulling over other ideas for finding a new source of income to augment his government pension, easing the financial pinch that had become a daily worry for a family man with so many children. He wanted to fix up the house a bit. Right now, even the most mundane roof repair seemed like an extravagance.

When Hanssen came to a small wooden footbridge that crossed over a meandering stream called Wolftrap Creek, he reached under the left corner of the structure and slipped an inch-thick package wrapped in a black, taped-up plastic bag over a rusted beam where dirt had eroded from the bank. The bag held a collection of seven FBI documents, classified SECRET, spelling out details of current FBI investigations against Russian spies. There was also a computer diskette containing an encrypted letter that read as follows:

Dear Friends,

I thank you for your assistance these many years. It seems, however, that my greatest utility to you has come to an end, and it is time to seclude myself from active service.

Since communicating last, and one wonders if because of it, I have been promoted to a higher do-nothing senior executive outside of regular access to information within the counterintelligence program. Furthermore, I believe I have detected repeating bursting radio signal emanations from my vehicle. I have not found their sources, but as you wisely do, I will leave this alone, for knowledge of their existance [sic] is sufficient. Amusing the games children play. In this, I strongly suspect you should have concerns for the integrity of your compartment concerning knowledge of my efforts on your behalf. Something has aroused the sleeping tiger. Perhaps you know better than I.

Life is full of its ups and downs.

My hope is that if you respond to this constant-conditions-of-connection message, you will have provided some sufficient means of re-contact besides it. If not, I will be in contact next year, same time, same place. Perhaps the correlation of forces and circumstances then will have improved.

Your friend,

Ramon Garcia

When he was sure the package was not visible to passersby, Hanssen turned and briskly retraced his steps. The walk took about five minutes. All the while, members of the Gees (short for SSG, or

Special Surveillance Group), non-agent personnel adept at blending into any environment, lurked in the bushes. They had not been expecting Hanssen this early. In his Palm Pilot, which FBI agents had covertly examined in advance, he had recorded: "ELLIS: 8:00." ELLIS was the code-name the Russians had chosen for the dead drop. The Gees, who had been following Hanssen day and night for a dozen weeks now, were ready. Just as he approached his car, a white Suburban van whipped around the corner and two tall, muscular young agents in "raid jackets," with "FBI" stamped in bright yellow letters front and back, jumped out, followed by two SWAT-team agents toting Heckler and Koch MP5 submachine guns — counterterrorist weapons — in case Hanssen reached for his gun. Other FBI agents and Gees emerged from cars and the woods, ready to block any effort Hanssen made to escape through the thicket. "Mr. Hanssen, you're under arrest," one agent declared, then recited a Miranda warning. The agents patted Hanssen down, handcuffed his wrists behind his back, and put him in the back of the vehicle, where two older case agents waited. A confidant of the Hanssen family said that when FBI agents slapped on the cuffs, Hanssen looked them in the eye and quipped, "What took you guys so long?"

The agents who drove Hanssen to an FBI field office suite in Tysons Corner, Virginia, said he had seemed resigned, that he had said little as they cuffed him. Once inside the field office, the agents laid out some of their evidence and played Hanssen a 1986 tape of himself speaking to a KGB agent.

Hanssen asked for a lawyer.

Ed Shaughnessy was an eighty-four-year-old neighbor who lived across the street from Foxstone Park. A little after four P.M., Shaughnessy had tried to back out of his driveway onto Fairway Street, a quiet slip of a road that ends precisely at Foxstone Park. He was surprised to see the street had been blocked off by a large, unmarked van and that cars were being forced to turn around. "I got out to see what was going on and found more FBI agents than I had ever seen in my life," Shaughnessy recalled. "There had to be at least twenty-five of them." Shaughnessy saw a big man approaching a gray Taurus, shoved against the car and handcuffed. He would surely have recognized Bob Hanssen had Hanssen's back not been turned: Shaughnessy saw him every weekday morning at 6:30 mass at their local parish church, Our Lady of Good Counsel. "He always sat in the back on the right by himself," Shaughnessy recalled.

Not far away, in Arlington, other Gees were watching a second drop site code-named LEWIS to see if Russian intelligence officers showed up to reclaim a package that had been left for Hanssen. The Gees had already opened it, counted the money inside — $50,000 in used hundred-dollar bills — taken it to the FBI lab to be photographed and analyzed for fingerprints, then carefully replaced it. Once they received the radio call that Hanssen had been arrested and would not be showing up to collect the cash, they seized it as evidence.

The Hanssens lived in the modest, middle-class community of Vienna, Virginia, about seventeen miles from Washington. Their home, purchased in 1987 for $197,095, is situated on a quiet cul-de-sac at the end of Talisman Drive, a neighborhood of neatly kept houses with basketball hoops and carefully edged lawns that fade into a small woods where children play after school. The headquarters of the Central Intelligence Agency is only a few miles away, and it is not uncommon in the community for friends and acquaintances to have jobs related to the intelligence field. Neighbors know to be discreet and generally ask few questions.

At the time of their father's arrest, Bob Hanssen's three sons and three daughters ranged in age from fifteen to almost thirty. Only the two youngest, Lisa and Greg, still lived at home, but all of the children were well known in the neighborhood, with lots of friends who had grown up with them. They all made an effort to be home for the neighborhood block party held each year on Memorial Day.

The Hanssens were respected parents in the private Catholic prep schools that their children have attended: the Heights, a boys' school for grades three to twelve, located in Potomac, Maryland; and Oakcrest, a small girls' school for grades seven to twelve, now located in McLean, Virginia. Barbara Falk, director of Oakcrest, considered Bob Hanssen a perfectly lovely man, the kind of person who would think to fetch her a glass of wine after a drawn-out PTA meeting when she was still surrounded by parents. "If you asked who was the most chivalrous man I ever met, I would say Bob Hanssen," Falk declared. "He is the consummate gentleman. I never even heard him swear."

Falk called the Hanssens an exemplary couple. "In my whole life, I have never seen a better marriage," she said. "Bonnie loves him madly. And I've never seen six kids like theirs. And if you see the children, you will see what successful parents they are."

As they were growing up, the Hanssen children often played street hockey on warm summer nights with other kids in the neighborhood, finishing only when it grew dark. Because no one had an ample backyard, the kids would gather at the end of Talisman, the cul-de-sac eventually taking on the air of a festive town square. Bonnie Hanssen enjoyed these carefree evenings and often brought her grandchildren in a stroller. The children's friends always felt welcome at the Hanssen house. The family's rectangular dinner table had a chair at each end and a bench on either side for the kids. The flowered tablecloths were always fresh and ironed. It was such a friendly neighborhood that the longtime principal at the Wolftrap Elementary School at the end of the block once said that she imagined that when she died and went to heaven, it would be like living on Talisman Drive.

Though the Hanssens had lived in the same community for almost fifteen years, no one was particularly close to Bob. He was considered a loner, an awkward man who discouraged small talk by averting his eyes to passersby, a man who seemed to prefer to take solitary walks with his dog than to chat on the street corner with neighbors. He spoke in a low voice, so people found they had to lean forward a bit to hear him. "I don't think anybody here knew him well at all," Diane Dondershine, who lives at the other end of Talisman Drive, told the *Washington Post*. Jerry Garrett, who lives next door, agreed: "The rest of the family is outgoing and friendly, but he never seemed to want to talk to anyone."

Nancy Cullen, who has known the family for fifteen years, often saw Hanssen on her morning walks. He parked his car on the street. His hours were extremely regular. By all appearances, he was just another guy going to work each day, fighting the good fight and all that patriotic stuff.

It's not that Hanssen was antisocial. Not exactly, anyway. He attended neighborhood events, made the right small talk, and never had much to drink, but his focus was always on Bonnie. At fifty-four, she was still beautiful, with shoulder-length auburn hair, clear porcelain skin, and large, soft brown eyes that always seemed to suggest she was thinking of something else. She was petite and well proportioned, with that same schoolgirl figure that had attracted him in the first place. Bob thought she looked a lot like the sexy actress Catherine Zeta-Jones. He loved the fact that other men still found her attractive. "You could always see that Bob adored his wife," Cullen said. "He was besotted by her. He would just gaze at her." Bob also loved his kids. When they were little, he read them bedtime stories, and as they grew older, he made a point of trying to attend their soccer games. He took particular pride in their academic achievements, often bragging about them to the guys at work. (They were all smart — like their father, of course.) He drew them into conversations about ethics at the dinner table, but he also took it in stride when they teased him, usually about his love of computers. He was always showing off new gadgets. But Bob was not one to engage other parents in conversation. "You approached him; he never approached you," said Cullen, an executive with the National Association of Broadcasters. "If you didn't say 'Hi' to him first, he might not even notice you."

At family gatherings — a recent Thanksgiving in Chicago, the wedding of his daughter, Jane — Hanssen often seemed detached, as if he were there yet not there at all. Gathered in Rome in 1999 for the ordination of Bonnie's brother, John Wauck, the entire family posed for a picture in front of a church. Hanssen stood to one side, his arm protectively around the attractive woman next to him, as if she were his wife. But she wasn't; the woman was Jeanne Beglis, Bonnie's sister.

Bonnie was standing behind Jeanne in the back row, barely visible. *It was as if Hanssen had forgotten where he belonged.*

In sharp contrast to her husband, Bonnie Hanssen was liked by everyone who knew her. Neighbors, friends, and colleagues all described her as an optimistic woman, sunny by nature, and with a good sense of humor. Despite her devout Catholicism, she could laugh at an off-color joke and lapse into the occasional profanity. Everyone knows married couples in which one partner is good company, the other a dud. With the Hanssens, Bonnie was the fun one, Bob the bore, always turning the conversation to the Linux operating system or religion or abortion politics and then putting everyone to sleep.

In fact, Bonnie had a moralistic streak in her, too; not as strong as Bob's, but it was there. She liked people to think she was beyond reproach, that she did everything according to the Good Book, that she was above sin. And she always talked about Perfect Bob, how smart he was, how hard he worked, what high ethical standards he had. She said these things even though she knew they hadn't always been true, that at least once he had been tempted by the devil. Maintaining the façade wasn't always easy; it took a nightly dose of Nyquil to get Bonnie to sleep. And many times when Bonnie smiled, she didn't really look happy. Her face seemed to reveal that as hard as they tried to be good Catholics, as often as they went to mass, as much as they followed the rules, she never thought they really measured up. Nobody ever does, of course, but it seemed to weigh on Bonnie.

She and Bob could also be very judgmental, always reinforcing each other about the right choices they had made in life while criticizing others, from politicians to neighbors, sometimes with rather unchristian remarks. Driving to St. Catherine's on Sunday mornings, Bonnie would complain about people who were jogging instead of being in church. But this was the private Bonnie, a side that most people never saw. Among the women in the neighborhood she was much admired — especially for the ease with which she seemed to raise her six children. The neighbors would ask themselves why *they* couldn't seem to bring their families together for a big pot roast dinner on Sunday evenings, why *they* couldn't turn themselves into the model wife and mother that Bonnie seemed to be. "And most of us did not have six kids," noted Nancy Cullen. Bonnie somehow managed to be frugal without seeming to sacrifice.

But this Sunday was different. When Bob failed to return home by dusk, Bonnie grew worried. He had not been looking after himself physically in the past year, and his weight had increased. He hadn't really been himself for months. He didn't seem to engage in conversation. Sometimes he just sat in a chair, looking morose. Bonnie was concerned about what would happen if Bob didn't take his heart medication on time. He also suffered from kidney problems and complained to friends about what a nuisance it was to have to get up in the middle of the night to pee. As darkness fell, Bonnie called her mother, Fran, and her sister, Jeanne, and asked them to pray for Bob. Finally, Bonnie could stand it no longer. She got in her van and drove to Dulles. A friend went with her for moral support.

What Bonnie did not know was that FBI personnel had been watching their movements from inside a house across the street. The government had secretly purchased the property almost three months before, just after Thanksgiving, not long after they put Hanssen under surveillance. It had been put on the market and sold abruptly. The couple who owned it had no children and moved away without even saying good-bye. For the past six weeks the blinds had been pulled tight and folks were beginning to wonder why no one moved in. Nor did anyone ever see people come or go. One neighbor said she should have realized something was up when a lineman remarked that the phone company was installing eight new lines.

The FBI may have fooled some of the neighbors, but had they fooled him? Neighbors say that after the telephone company put the lines in, the FBI failed to remove all those little red flags that the utility company used to mark the gas lines to the house. (Talk about a red flag!) Did they think Hanssen wouldn't be suspicious when the house directly across the street sold and yet appeared uninhabited? Not likely.

Another time Hanssen noticed that when he drove into the FBI parking lot, something in his car made a distinctive "ping." Was there a tracking device attached to his vehicle to monitor his movement? Well, yes, there was; and the FBI put it there. But bureau officials insisted after the arrest that their device would not have made any noise. If the ping existed, they said, it did so only in Hanssen's increasingly paranoid imagination.

Hanssen had also believed that unusual transmission bursts were affecting his new, Global Positioning System, the electronic navigation device that pinpoints one's exact position on the earth via satellite. Nothing pleased him more than to be driving north and having his GPS say he was driving north. But every few minutes it would fly off kilter and then correct itself. He knew something unusual had to be causing it, that he was probably under surveillance. He assumed the transmitter that the FBI was using to track him was somehow interfering with his GPS unit. It pissed him off. On the day before his arrest, someone in the neighborhood saw Hanssen marching up and down his street for twenty minutes, directly in front of the house purchased by the FBI, holding his new yellow GPS gadget and watching for transmission bursts. Later, the onlooker wondered if he was giving the FBI the finger.

Some wondered if Hanssen really did know what was up, or whether his claims were simply boasts, attempts to show that he had once more outsmarted his colleagues — the FBI had not caught him, he had caught himself. He had willingly walked into the arms of the arrest team because he wanted his years of espionage to be over. But if that were true, FBI officials said, why hadn't he just walked into the field office and given himself up? Why put his wife and children through the shame of a public arrest near their home?

The FBI certainly made its share of mistakes, but its agents did know how to tail a target. Two Gees followed Bonnie to the airport. When she pulled into the parking lot, they approached her, identified themselves, and told her that her husband was fine, he had not been hurt. But there *was* a problem — an agent arrived next and told Bonnie he had some bad news. Would she accompany him inside the airport, where they could talk privately? Bonnie nodded numbly and followed the agent to a room borrowed from airport management. There he broke the news to her: her husband of thirty-three years was under arrest for espionage. When the agent was satisfied that Bonnie's shock was sincere and unrehearsed, her bewilderment too deep to be feigned, he drove her home.

Bonnie Hanssen was not under suspicion. FBI officials considered her and her six children victims. But in case she had any useful information to pass on — and for the family's physical safety, as well — two agents spent the night in the Hanssen home. They listened as she telephoned relative after relative, including Bob's eighty-eight-year-old mother in Venice, Florida. At some point, Bonnie was permitted to speak to her husband. Their conversation lasted less than a minute. Hanssen told his wife that he was very sorry and would always love her.

But now came the hardest part. How in the sweet name of Jesus was she going to break this news to the children? It would be easier to explain if he had died in a traffic accident — tragic, of course, but those things happen. In many ways, this was worse. All of a sudden, like a brutal whack in the head, life as they knew it would be forever changed. Daddy, the tall, strong voice of morality and authority, the guy who set the example, laid down the rules, got them all in the van for Sunday mass . . . in jail, for the rest of their lives? It was unimaginable, a nightmare. Surely Bonnie would wake up in the morning, tell Bob about this wretched dream, and they would shake their heads the way married couples do, wondering what it was they ate for dinner that could have caused such a terrible, awful night.

But there was no shaking it.

The next day, realizing that the house would soon be surrounded by television trucks and curiosity-seekers, the FBI offered to put Bonnie, Lisa, and Greg up in a hotel for a few days. They accepted.

The voices of Bonnie's friends said it all.

Later, those who had the courage to talk to her spoke not in the warm tones of sympathy reserved for death and dying, but with words of pity (Oh, poor Bonnie . . .), those well-meaning but sickening comments that are hardest of all to bear. Bonnie, the woman who never drew attention to herself, who reveled in anonymity, was suddenly at center stage. And a hushed, if hackneyed, line of inquiry began to be heard: *What did she know and when did she know it?*

Everyone agreed on one point. Bonnie's burden — physically, emotionally, financially — was staggering, so much so that no one was sure that the enormity of the events had truly sunk in for her. Finally, it was one of the boys who explained it all to her by reading out loud the allegations against their father laid out in the FBI's 109-page affidavit. The extraordinarily detailed document stated that Hanssen had given up invaluable secrets to the intelligence services of what used to be called the Soviet Union and is now Russia. He had also compromised four Russians who had been recruited as spies for the United States. Three were recalled to the Soviet Union and executed.

In page after page of stunning detail, the government accused Hanssen of giving the Soviets upwards of six thousand pages of classified material and twenty-six computer disks detailing the bureau's "sources and methods," including its latest techniques for electronic eavesdropping. It charged also that Hanssen turned over classified reports describing U.S. defenses against nuclear attack and other aspects of U.S. intelligence, and that he sabotaged the FBI's investigation of former State Department employee Felix Bloch in 1989. Hanssen was charged with revealing a breathtaking array of sensitive human and electronic sources within the Soviet government and its successor state, the Russian Federation. This was not only the worst blotch on the FBI's record, it was one of the most damaging penetrations of the American national security apparatus in history.

Bonnie listened intently as her son read the details. Then she began to pray.

At 6:30 mass on Monday, February 19, at Our Lady of Good Counsel in Vienna, Ed Shaughnessy noticed that Bob Hanssen was not in his usual place. He thought back to the arrest he had observed the day before — Shaughnessy had assumed it was a drug bust — and wondered if the FBI was working its agents harder than usual.

The next morning Shaughnessy noticed that Hanssen was absent again. "I thought maybe the FBI summoned in all the agents after the drug bust," he recalled. "That must be why he wasn't there."

In the Hanssen neighborhood, too, there were signs of something unusual. Anyone who looked out the window saw more cars than usual lining the street. They all had District of Columbia license plates. A few people wondered why they were there.

Meanwhile, in the Alexandria courthouse, a silent Robert Hanssen was being arraigned. His defense attorney turned out to be the legendary Plato Cacheris, who several years before had successfully negotiated a plea bargain agreement for CIA spy Aldrich Ames that spared his wife a long jail term. Other lawyers envied Cacheris's press clippings; *Time* magazine once observed that he "cross-examines with laser-like ferocity and charms the jury with wit. 'My client is a fool, an ass, a boor,' he once thundered. 'But he is not a cold-blooded strangler.'" Cacheris had built a tennis court in his backyard with fees paid by Attorney General John Mitchell, whom Cacheris defended during Watergate. He defended Fawn Hall, a key figure in the Iran-Contra scandal a decade before. Most recently, he was retained by White House intern Monica Lewinsky after her sordid affair with President Clinton.

In downtown Washington, on the seventh floor of the Hoover Building on Pennsylvania Avenue, FBI director Louis Freeh was steeling himself to announce the worst espionage scandal in bureau history. Hanssen was only the third FBI agent to be accused of spying, and he had been spying longer than any of the rest. It would take years just to calculate the devastation.

In some cases in which a U.S. government employee sold secrets to the KGB, there had been telltale signs, "risk factors" in the parlance of the intelligence community. CIA officers David Barnett (arrested in 1980), Edward Lee Howard (who defected to the U.S.S.R. in 1985), and Harold Nicholson (arrested in 1996) had, among them, career, money, and drinking problems. The most damaging CIA turncoat, Aldrich Ames, had, as they say, all of the above. John Walker, a retired Navy warrant officer, had been blatantly greedy. FBI agent Richard Miller (arrested in 1983) had been in trouble with the bureau for years. The only case nearly as perplexing as Hanssen's had been that of FBI agent Earl Pitts (arrested for espionage in 1996). But unlike Hanssen, he had neither children nor church to keep him straight.

The Hanssen case was different. Bob Hanssen seemed to be a humble and pious man, awkward and ill at ease perhaps, but generally a good citizen. Those who knew him say there was no doubt that he adored his wife, children, and four grandchildren, and that they loved him in return. But he kept all others at a distance.

"After the service, I always stand outside the church to greet people, and he never came over to say hello," said the Reverend John M. O'Neill, pastor of Our Lady of Good Counsel, where Hanssen attended the daily morning mass for at least ten years. "He exited out the side door." Two other priests at churches Hanssen attended regularly for many years say they knew him only slightly.

Later, after the news of the arrest had spread, Father O'Neill realized that his parishioner looked at spying as a game, that he lived in a world of shadows and secrets and deception that few people ever even imagine, much less know. "He was dancing with fire," O'Neill said upon reflection. "What is threatening about fire is also its thrill. This is a theme that would have developed earlier in his life."

And so it did.

2

EARLY YEARS

Robert Philip Hanssen was born April 18, 1944, the only child of Howard and Vivian Hanssen. They called him Bobby. The family lived at 6215 North Neva, a quiet avenue lined with towering Dutch elms, located in Norwood Park, on the western edge of Chicago. In many ways, the village resembled Chesterton's Saffron Park, where "the pretensions to be a pleasant place were quite indisputable." Norwood Park was a modest, working-class neighborhood of brick cottages and wooden bungalows.

The Hanssens lived in a narrow, white frame house with a high dormer window. Built on a small, odd-shaped lot, it had two bedrooms downstairs, connected by a bathroom. Bobby's paternal grandmother lived on the second floor in a makeshift apartment, and the backyard was fenced to keep track of the family's black cocker spaniel. Ruth Kremski, who lives next door, sometimes talked to Howard Hanssen over the fence, and Kremski considered the Hanssens nice neighbors, "quiet and friendly." Warren Peterson, who lives on the other side of the Hanssen house, called the family "wonderful people." Once when Peterson's car was stolen from in front of the house, Howard Hanssen drove him to the police station and helped him with the paperwork.

In Norwood Park there were no African-Americans or Asians, and only two Jewish families. Most residents attended either the United Methodist Church or Our Savior's Evangelical Lutheran Church, which were directly across the street from each other. Vivian Hanssen was a Lutheran and walked around the corner to church almost every Sunday. Howard was not as committed a churchgoer, and nobody remembers their son as being particularly religious.

The big event of the summer was an annual international bicycle race in Norwood Park Circle which, lore has it, used to be a race track. Cyclists came from all over the world to participate in the all-day race, and the circle, which curves around the neighborhood, would be lined with spectators. Many neighbors had ice cream socials to celebrate. The community also had its share of characters. When the Caitis family moved into their white house with green shutters, the local piano teacher walked into their living room, sat down at the piano, and sang a little ditty she had composed called "Welcome to the Neighborhood." It was a community proud of itself and its spirit.

Howard Hanssen, who was born in May 1911, had been a Navy petty officer during World War II, assigned to patrol boats monitoring U.S. shores. When his duty was over, he returned to Chicago and joined the Chicago Police Department on January 9, 1943. It was not a department — or, for that matter, a city — known for hewing to the straight and narrow. The late *Chicago Tribune* columnist Mike Royko summed up the situation like this: "Most Chicagoans considered the dishonesty of the police as part of the natural environment."

During a career of almost thirty years Howard Hanssen rose through the ranks of the Chicago Police Department. Most of that time he was assigned to District 16, the police precinct in his neighborhood, which was not far from Chicago's O'Hare International Airport. According to John O'Brien, a former *Chicago Tribune* reporter and author of several books on Chicago crime, Howard Hanssen was regarded by colleagues as bright and honest. In the early 1950s, Hanssen was promoted to sergeant. Retired Chicago police detective Norb Handley told O'Brien that he recalled working with Hanssen in the mid-fifties in the old 39th District traffic unit on North Damen Avenue. "I knew Hanssen as a gentleman who was not on the take," Handley told O'Brien.

Hanssen's true passion was for horses. He even owned a share of a racehorse in Sebring, Florida, and often showed off pictures of it to friends on the night shift. Steve Belak, now a retired Chicago police detective, used to tease Hanssen with the same old joke: "Why," he would ask, "is there no horse manure out on the racetracks? Because all the horses' asses are in the stands." Belak thought the crack pretty funny and always laughed. Hanssen, a rather humorless man, never did. "He'd just look at me with this blank stare," Belak told O'Brien.

In 1961, at the age of fifty, Howard Hanssen made lieutenant. In 1970, two years before his retirement, he was transferred to Unit 135, the Intelligence Division, which divided its staff between the old Maxwell Street Station and the department's downtown headquarters on South State Street. He was the lieutenant in charge of the division's eighth-floor file room, which contained thousands of sensitive and secret documents. Their subjects ranged from organized-crime figures to high-end burglars, strong-arm robbers, major career criminals, and radicals of various stripes. They also included people who had committed no offense but to criticize City Hall. "This was not a high-profile post, but it was certainly one of importance," said John O'Brien.

The fear of communism had by then been building in the United States for over two decades. In February 1948, when the Soviet Union seized the tiny, democratic country of Czechoslovakia, Americans began worrying about war with Moscow. In September 1949, President Truman announced that Soviet spies had stolen the secret of the atom bomb. In January 1950, Alger Hiss was convicted of perjury amid charges that he spied for the Soviet Union while at the State Department in the 1930s. The next month, Senator Joseph R. McCarthy, an obscure alcoholic Republican from Appleton, Wisconsin, claimed in a speech to have a list of 205 State Department employees who were members of the Communist Party. Although

McCarthy produced no evidence to support the allegations, the Washington press corps (with few exceptions) rushed to print his charges without questioning their accuracy. Many Americans were convinced there was "an enemy within" working on behalf of the Soviet Union. The fear was reinforced when Julius and Ethel Rosenberg were executed for treason in 1953.

It was not entirely unwarranted. The number of Soviet intelligence officers assigned to the United States had been steadily growing since the end of the war. Americans worried that Communist agents were bribing politicians, taking over labor unions, and infiltrating the armed forces and police.

In 1970, America was no longer in the midst of a McCarthy-like Red scare, but anticommunism was still on the agenda. Richard Nixon, who had been credited with having "got" Hiss, was president. And in Chicago, a city crowded with families rooted to Central European nations now a part of the Warsaw Pact, communism was still a live issue. The city's Red Squad, which was a unit of the Intelligence Division of the Chicago police, tracked subversives and kept dossiers on more than 250,000 private individuals and lawful organizations until it was sued for First Amendment violations and was quietly disbanded in 1975. But when Robert Hanssen's father joined Unit 135, the hunt was still afoot. Some Norwood Park neighbors insisted that Howard Hanssen was a member of the Red Squad, but no one knew for sure.

Bob's mother, Vivian Hanssen, was a quiet, gentle woman. Whether she ever wanted more children was not a question anyone in the neighborhood would have asked, certainly not then, anyway. Friends describe her only as reserved and very protective. But her support for Bob could not totally compensate for her husband's disappointment in his son.

Because there were alcohol-related problems in his family, Howard Hanssen was a nondrinker, but he had a manipulative personality that often made him difficult to be around. "He was a dry drunk, very controlling," said a Hanssen family friend. A harsh taskmaster, he was extremely strict with Bob, ordering him about with military harshness. Family friends recall a boy fearful of his father, though they doubt that Howard ever beat his son. But, said a classmate, "There is no question that Bob was verbally abused. . . . His father had a real mean streak." The elder Hanssen customarily spoke about his son in disparaging terms to other parents, wondering aloud whether his child would ever amount to anything.

Howard had great ambitions for Bob. The veteran cop wanted his son to become a doctor or dentist. Both seemed possible; while he was too awkward and insecure to ever be the teacher's pet, Bob was a good student when he wanted to be. "He didn't raise his hand a lot, but we sensed he was smart," said Karen Caitis Lison, a schoolmate. "He always looked a little bored in school, but when the teacher would announce the highest grade, it was often Bob. He never made a big deal of it."

On the surface, Bob never made a big deal about *anything*. But he didn't like surprises, and he didn't like feeling forgotten. On his first day of school at the Norwood Park Elementary School, he burst into tears when he didn't find his mother waiting for him at the school-house door as she had promised. In fact, she was at the door, but it was the outer door, not the first one that Bob opened. "When he didn't see her, he totally lost it," a friend recalled.

Classmates remembered Bob being tall for his age and so thin that his chest seemed somewhat sunken. His teeth protruded slightly, and he often had a bit of drool in the corners of his mouth. He also dressed differently from the other boys, who thought it "cool" to wear solid-colored shirts — Bob usually had on thin plaid ones chosen by his mother. Also, most of the boys wore low white gym shoes while Bobby had blue ones. From afar, these differences seem trivial. But at an age when conformity is the norm, Hanssen's dress stood out.

The thick, white belts worn by student members of the school safety patrol were a sign of status and responsibility. One year Bobby was captain of the patrol, though he was not a popular leader. If other children were not on their assigned corners, Bobby would tell on them to the gym teacher, Mr. Valone, who was in charge. "No one else ever did that," said Karen Lison.

On rainy days, the children were not permitted to go out on the grassy playground for recess. Instead, they played a game in which one child would run up and down the aisles of the classroom. When the child squeezed into another student's chair, the child sitting down had to get up and run. One day Lison felt sorry for Bob because no one ever sat in his seat. So she decided she would sit down next to him. "I got the funniest look from him," she said, "this feeling of someone saying, 'I don't need that.'" Lison said she never again tried to include Bob in activities: "It didn't seem to bother him to be a nerd. He had this sense of self-importance, almost as if he liked to be uncool."

Bravado aside, no child wants to be left out. All kids experience a sense of exclusion at some time or other that sets them apart, but one event really did make Hanssen different from the other kids at school. When Bobby was in sixth grade, his closest friend, a short blond boy with a round face named Paul Steinbachner, was hit by a ball in a game at school and knocked down. A somewhat frail child, Paul was proud not to get a nosebleed. The boys liked to build models together, and Bobby accompanied Paul home that afternoon after school. As former classmates tell the story, the boys were playing in the house and Paul complained of a headache. He went into the bathroom. A few minutes later, Bobby heard a thud. Paul had died, apparently of a brain hemorrhage.

For most children in the neighborhood, this was their first exposure to death. The entire class attended a wake for Paul. Yet, in school the next day, there was hardly any discussion of the incident. "We'll all miss Paul," Mrs. Beemer, their teacher, told the children. Then she asked them to bow their heads. "That was it," said Karen Lison, now a psychotherapist. "Nothing more was ever said." For Hanssen, it probably felt like a punch in the gut; his best friend had died, virtually in front of him, and nobody seemed to care — not about Paul, and not about him. How the incident affected Hanssen is unclear. Perhaps it caused him to withdraw further, scared of or angry at the world. But one thing is certain: in more than a dozen interviews with Hanssen's former Norwood Park classmates, almost all pointed to the death of Paul Steinbachner as having a profound impact on Bob's personality as a child, and perhaps even as an adult.

In 1958, Bob enrolled in William Howard Taft High School. It was a large school in an imposing red brick building, located at the corner of Natoma and West Hurbut, about a mile from the Hanssen home. From the playground, students could see the Northwest Highway, a major artery that was just being built. Taft drew its student body from several nearby elementary schools and was considered one of the best public schools in the city, almost on a par with those in the wealthier suburbs of Park Ridge and Glenview. Since its doors had first opened in 1939, enrollment had almost doubled to about three thousand students. Most Taft students took their education seriously: almost half the male graduates went to college, and the female graduates who did not go to college usually found a job to keep them busy until they married. Not surprisingly, football was the most popular sport. At Saturday home games, the bleachers were filled with parents and students waving pompoms of blue and silver, the school colors, and screaming themselves hoarse.

After the Soviet Union astonished the world by launching *Sputnik,* the first man-made earth satellite, on October 4, 1957, high schools and colleges throughout the United States were challenged to improve their math and science programs, in order to catch up with the Russians. Taft was no exception. One of its strongest math and science students was Bob Hanssen. He was the first student Karen Lison ever saw with a slide rule, which he wore in his shirt pocket. Bob usually placed into the advanced, or "accelerated" classes. The city had a requirement that high school freshmen take a general science course. But the school system had recently implemented a test in eighth grade that, if passed, allowed students to go directly into biology, which was considered a second-year course. Bob and his new friend Jack Hoschouer both passed the test. Sophomore year Bob took physics, and as a junior, he and Jack became lab partners in chemistry class. "We used to get dry ice and put it in a flask with Coke and pretend to drink it," Hoschouer recalled. In the lab, pretending, Hanssen seemed to have found his element.

There were many after-school activities available to Taft students. Drama Club and Band were popular. So were the Mixed Chorus and the Dance Committee. Hanssen chose to join only the Honor Club, which required a 3.5 grade-point average, and the Radio Club, a small group of boys who built ham radios from mail-order kits. To acquire an operator's license, applicants had to pass tests in radio theory, regulations, and Morse code. "It was difficult to do compared to the tests today," said Tom Kozel, a fellow ham radio operator, who was also strong in math and science. Once licensed, they could communicate by radio with other hams around the globe, a forerunner of the Internet. "I would equate our interest in ham radios with the interest kids have today in computers," Kozel said.

After school, the boys would go to each other's houses for dinner and to spend the night, poring over Heathkit catalogues and fooling around into the wee hours with their radio sets. In the midst of the Cold War, and just over a decade since the end of World War II, it's hard to imagine that at some point Hanssen and the others didn't fantasize that they were engaged in more than casual radio chatter. Up late, their activities their own little world, their conversations transmitted only to those who knew their call numbers — to a high school misfit, it was pretty close to being a secret agent. And like so much about Hanssen, these activities were not much discussed outside of a very intimate circle. "Whenever he would join a club, or later, in college, go out with a girl, he would probably tell you after rather than before," Kozel said. "In retrospect, you might say he was secretive, but it didn't seem that way at the time."

During these years the United States' grim rivalry with the Soviet Union continued to build. In early 1960, an American U-2, a supersecret, high-altitude reconnaissance plane was shot down over Soviet territory, and its pilot, Francis Gary Powers, was captured and sentenced to ten years in prison. It was a severe embarrassment for President Eisenhower, who had initially denied that such flights ever took place.

These stories were front-page news in the Chicago newspapers, but far removed from the world of Bob Hanssen and his friends, background noise for teenage boys who were more interested in cars than current events. When Bob and Jack got their driver's permits, they liked to cruise around town, supposedly looking for girls, but more likely content to gawk. Bob's father owned a pale blue Dodge, a late-fifties model with big tail fins, and the boys often took off together after school. "We hardly made any serious attempts to pick up girls," Hoschouer said. Instead, they cooked up pranks to play. One night they came out of a high school dance at ten-thirty or eleven, and Jack drove his car down the sidewalk rather than in the street. The chemistry teacher stood there watching them, just scratching his head. "This was as wild as we got," Hoschouer said.

On weekends in June they would sometimes go to the "Road America" sprints in Elkhart Lake, Wisconsin. Elkhart Lake was the midwestern Mecca of road racing, a magnet for college students and car enthusiasts, located on a beautiful parklike setting on the edge of Wisconsin's Kettle Moraine State Forest. The circuit was four miles around and included fourteen turns. As Jim Donick, editor of *Village Sports Car,* described the scene: "Spectators could stand at the foot of the pagoda on the front straight, where they might watch the cars breaking hard from 140 MPH at the bottom, then sliding through 30 MPH Turn 5 to go screaming up the hill and under the bridge." The local brew was Leinenkugel beer. "We would eat bratwurst and look at girls, eat bratwurst and look at girls," Hoschouer recalled.

By their senior year, the school superintendent had organized a college-level course for the best science students in the city's three high schools. It was taught by Lorraine Truzynsk, who was working on her Ph.D. Five students from Taft were selected; Bob Hanssen was one of them. The class was held at Chicago's Von Steuben High School, about five miles from Taft.

Every day at one P.M., the boys would pile into an old car owned by one of the five science students. The aging Studebaker was coated with a battleship gray primer and was powered by an often temperamental engine. The trip usually took forty-five minutes, not counting cold winter days when the fuel line would freeze up and they had to push the car to a garage. "I got to know Bob pretty well," said John Lorenz, one of the five and now a psychiatrist at Evansville State Hospital in Indiana. Lorenz remembered Hanssen as a mediocre student. "Bob was pretty smart, but a little on the lazy side," Lorenz said. "If he could think of an easy way to do something, he would. In solving problems, if something took a lot of time, he wouldn't do it. He did the things he could do easily. To some degree high school seniors can all be that way, to look for the easy way to do things. But this was a little remarkable."

It was not until their senior year that Bob and his friends began to date. At the fund-raising carnival for Resurrection Hospital one summer, he won a teddy bear at the shooting gallery and waved it in front of a pretty girl who was walking by. Her name was Paulette. "They were a hot number for a while," Hoschouer said. Like most high school romances, it didn't last, but it was a big step forward for Hanssen, whose behavior around women usually oscillated between sullen introversion and misguided cockiness. One day he decided that the boys should drive to Rainbow Beach, which was on the South Side of Chicago, a good hour's drive away. Jack countered that they should go to Hollywood Beach or North Avenue Beach, both considerably closer. But Bob was adamant. When they arrived and stretched out on the sand, they noticed a couple of girls nearby. Some teenage boys were showing off in front of them. Bob got up and went over to ask the girls if the boys were bothering them. They said no, but the show-offs left anyway. The girls seemed disappointed. Bob returned to the blanket pleased with himself, convinced he had rescued them. "I think he had a damsel-in-distress syndrome," Hoschouer said.

As a way to celebrate their graduation that summer of '62, four of the boys decided to go backpacking in Colorado. By fall, they would be going in different directions, and they were eager for one last good time together. They spent weeks planning the trip; to avoid asking their parents for a car — few families at the time had two — they planned to take a Greyhound bus. When it was time to make the final arrangements, all the parents gave their consent, except Vivian and Howard Hanssen. "Howard liked to fuck with Bob," said a friend. "It was just like him to let him plan that trip and then not let him go."

Bob ended high school just as he began it: left out.

3

KNOX COLLEGE

Bob Hanssen and ham radio buddy Tom Kozel arrived, with 349 other freshmen, at Knox College on Thursday, September 13, 1962. The small, liberal arts school is located in Galesburg, Illinois, the regional center of commerce and social life in an isolated slice of prairie in the Mississippi River valley, some 200 miles southwest of Chicago.

Hanssen was assigned to the second floor of Rabb Hall, a two-storey red brick dormitory in the center of the Old Campus. Each floor housed twelve students, two to a room. "My mother recalls that when it was time for Bob's parents to leave, he did not want them to go," Kozel said.

Hank Wilkins, who was Bob's dormmate and friend for four years, remembers that when they met, Bob was "over six feet and lean, like a string bean." He was well dressed, neat, and had good manners. "Bob did not stand out," Wilkins said. "He was part of the pack. He was not showy. Bob liked to be a part of the group, but he didn't like to draw attention to himself." But while somewhat shy, he also seemed to have a penchant for telling small lies to impress his friends. Wilkins recalls

Hanssen telling him that his father was a major power in the Chicago Police Department and that he had put a lead door on their house so criminals could not shoot through it.

Founded by Presbyterians and Congregationalists in 1837, Knox College had a long reformist tradition that began with the then popular idea of combining manual labor and academic work. It had been a regular stop on the Underground Railroad for escaped slaves on their way from St. Louis to Chicago and Canada. In 1858, it became the center of national attention when the fifth of the Lincoln-Douglas debates — this one on the issue of slavery — was held on the campus.

When Bob Hanssen arrived at Knox, the college's academic reputation was at its peak. The school was also in good financial shape. A fine arts building had recently been built, and work was under way for a new math and science center. The student enrollment, which included a record number of National Merit scholars, had increased to slightly over a thousand. "This was one of the high points in the history of the college," said Owen Muelder (Knox '63) and now director of alumni relations.

The $2,350 fee for tuition, room, and board was considerably higher than the cost of attending state schools, and competition for admission was fierce. (Hanssen had received a chemistry scholarship.) It was an atmosphere where students were expected to work hard and perform well. Of the freshmen in Hanssen's class, 77 percent had been in the upper quarter of their high school graduating class, and their median SAT scores were close to 1,200. After graduation, 62 percent of the men and 30 percent of the women would go on to pursue advanced degrees. "There was a robust intellectual life," said Robert Seibert, a political science professor at Knox, who himself graduated from the college in 1963, the year after Hanssen arrived.

Knox College was the first school in Illinois to teach Russian. The Cold War was at its peak, and as the fear of Soviet influence grew, many thought Russian could become the dominant foreign language of the next half-century. While there was some queasiness among Knox administrators about encouraging student interest in the

Soviet Union, science majors were encouraged to study Russian as part of their language requirement.

There was one Russian-language teacher at Knox, a Yugoslav Ph.D. who had come to the school in 1959. His name was Momcilo Rosic. An officer in the defeated Royal Yugoslav Army, he had spent part of World War II in a German prisoner-of-war camp. "I was very nationalistic," said Rosic. "I wanted nothing to do with communism." The dramatic story of Rosic's capture and escape, which he sometimes told in class, captivated his students.

The Russian curriculum consisted of two years of basic grammar and a third year concentrating on the nineteenth-century Russian writers Aleksandr Pushkin, Nikolai Gogol, and Ivan Turgenev. Rosic had no recollection of Bob Hanssen and didn't think Hanssen took more than two years of basic Russian. By the professor's standards, Hanssen could not have been much of a language student. Yet Hank Wilkins had a different impression of his roommate's deftness with languages. He said Hanssen liked showing off his German and Russian to his friends. "Most of us just did translating," Wilkins said, "but Bob could speak phrases. He would often throw out foreign phrases to us." To those who didn't know better, and few did, Hanssen gave the impression that he was an exceptionally smart guy.

Knox freshmen had a full selection of campus activities. A physics professor inaugurated an eight-to-ten-week seminar for students and faculty on computer programming. It was a new field, and twenty students signed up. Pulitzer Prize–winning poet Archibald MacLeish visited Knox to speak about the violence that attended the enrollment of James Meredith at the University of Mississippi, insisting that "the drama of Oxford was the fault of all of us." The theme of the 1962 homecoming in October was the Ringling Brothers and Barnum & Bailey Circus, and, as usual, the Knox football team, the "Siwashers," lost to its arch-rival, Monmouth College, with a score of 20–6. Greek life, with six fraternities and five sororities, was strong. In November, the Knox Theater produced Shakespeare's *The Taming of the Shrew* and the Gilbert and Sullivan operetta *The Mikado*. Comedienne Phyllis Diller also came to campus. In February, four months after the United States learned that the Russians had begun installing missiles in Cuba, *The Student* ran a page-one story on thirteen Knox students who were going to Washington, D.C., to demonstrate against the use of nuclear weapons. They picketed in front of the Soviet Embassy.

Hank Wilkins said that Bob was not part of any particular crowd and "would often disappear into the woodwork." When Wilkins and Hanssen could not find anything else to do, they would amuse themselves by playing softball on the quad next to their dorm. "On Saturday night, we would walk to the depot at two A.M. for eggs, toast, and hash," Wilkins said. "It cost fifty cents." The depot restaurant, known as the "Q," was named for the old Chicago-Burlington-Quincy Railroad that ran through town.

For Wilkins, now a retired high school math teacher, one memory of Hanssen stands out. It was the end of their freshman year at Knox. Bob was taking the Western civilization final. "He didn't like the first question, so he walked out of the exam and went to the gym," Wilkins recalled. "He spent two and a half hours working on his left-hand layup. That was Bob. Inner challenges were more important to him than those presented to him by others. If he had a challenge, he would put his heart and soul into it. Otherwise . . ."

In 1963, Kim Philby defected to the Soviet Union. Philby was a rising star in the British Secret Intelligence Service — and a Soviet agent. As Evan Thomas wrote in *The Very Best Men,* Philby was "the most effective 'deep penetration agent,' in the formal jargon of spying, in modern history." Recruited by Moscow in the 1930s, while a student at Cambridge University, he rose to the top ranks of the British civil service. A man of great charm, he enjoyed the company of Americans and was chosen to serve as liaison to the American intelligence community, assigned to Washington in the years 1949 through 1951. That position, wrote Thomas, "gave him access to American war plans against the Soviet Union (code name Trojan, a massive A-bombing campaign against 50 Russian cities), to American troop movements in the Korean War, to every single CIA operation against the Kremlin in the Baltics, Ukraine, and Albania." Philby transmitted secrets with encoding equipment hidden in his basement. It was something straight out of the movies — or the Taft Radio Club.

And say what you will about Philby, he certainly was noticed. Nobody, certainly not Bob Hanssen, could deny how clever Philby was. He had outsmarted them all.

John Lorenz, who had been Hanssen's science classmate at Taft High School, spent two years at the Illinois Institute of Technology before deciding he did not want to become an engineer. He transferred to Knox, where he ran into Bob soon after arriving on campus. Hanssen said he was planning to go to dental school.

Most Knox students got paying jobs in the summer, and Hanssen was no exception. For three summers, he worked at the Chicago State Mental Hospital. Each student was assigned to a ward where he or she was expected to interact with patients. If a patient said something irrational, the students were instructed to say, "Hey, you just said something crazy." Hanssen found the work interesting. There was another motivation to work there as well. "All the student nurses from the Chicago teaching hospitals went there for a summer to study," said Jack Hoschouer. "After one summer, he told me I ought to come work there instead of punching holes in metal parts for a metal manufacturing company. The pay wasn't great, about two hundred fifteen dollars a month, but the girls were pretty."

Indeed they were. At the hospital Jack Hoschouer met a nurse who would later become his wife. Bob got lucky, too. Bernadette Wauck, whom everyone knew as Bonnie, was a striking woman, petite with deep brown eyes and a narrow jaw. "She looked a lot like Natalie Wood," Hoschouer recalled. Bonnie had been raised in a Roman Catholic family of eight children in Park Ridge, Illinois, only a few miles from Norwood Park. Bonnie's father, Leroy Wauck, was a psychologist who taught at Loyola University, a Jesuit institution in Chicago. Bonnie's family was devoutly religious, and they often discussed their faith over the dinner table. Everyone in the family was bright, and their conversations often took an intellectual bent. Bonnie's mother, aunt, and two of her brothers were members of Opus Dei, a tiny, ultraconservative Catholic society that requires members to attend daily mass and regular confession.

Bonnie was a sociology major at Loyola when she met Bob Hanssen; her father had helped her get the job at the state mental hospital where Hanssen and Hoschouer were working for the summer. Bonnie had always been very popular, and lots of young men wanted to date her. Bob, however, was still somewhat awkward around women. He had dated a Chinese-American woman while at Knox, but his father had disapproved and the relationship ended. He initially showed his attraction for Bonnie somewhat oddly, by teasing her unmercifully about her driving skills. But the chemistry of love has no formula. For some reason, Bonnie wasn't offended. She had never felt quite as smart as her brothers and sisters, and she admired Bob's intelligence. His future looked bright. When he finished dental school, they would probably end up living in the Chicago suburbs — which is exactly what Bonnie wanted. She liked having her family nearby. Bob, in turn, worshiped Bonnie. At first, Jack Hoschouer thought she was a fluffhead, but he changed his mind quickly. Bob admired Bonnie's character. He liked her family. And he loved having a pretty, vivacious girl who met his handsome friend's approval.

Some nights the guys would hang out alone. They would venture down to Chicago's Old Town, a North Side entertainment district filled with hip bars, comedy clubs like Second City, and coffeehouses full of folksingers imitating Woody Guthrie. For kids from the suburbs, this was an exotic neighborhood. One evening they were walking down North Wells Street, and a young boy asked if they wanted their shoes shined. Hanssen picked the kid up by his shoulders. "He moved him out of his way like a piece of furniture," Hoschouer said. "Bob could be hard-headed." But it was not just that. He was increasingly annoyed with people standing in his way, literally and figuratively.

When Hanssen returned to Knox College for the start of his senior year, he began writing to Bonnie. She loved his letters, which were much more lyrical than he was in person. "She was impressed with his creativity, his imagination and his intellect," Hoschouer said. "She told me that she fell in love with his letters. Most girls thought he was weird."

Like all college seniors, Hanssen worried about his future. He applied to the Northwestern University School of Dentistry, which required applicants to take a dental aptitude exam to be admitted. John Lorenz said he didn't think that Hanssen was determined to become a dentist: "He never seemed that interested. Having been to medical school myself, I can tell you that you have to have the fire in the belly. You have to really want to be a doctor. If you really wanted to be a dentist, you would take hard chemistry classes and keep your grades high. You'd want to know about amalgams and how gold works. You'd think he'd want to know. But he seemed like he was kind of so-so about it."

Northwestern required dental-school candidates to show a high undergraduate grade-point average, usually a minimum of 3.2. In his four years at Knox, Hanssen never made Dean's List, which in his senior year required a 3.35 average. "I remember Bob telling me that he had been admitted to Northwestern, but that they said he would have to maintain a certain GPA in his senior year or they would kick him out," said John Lorenz. "It was a conditional admission." Later that year, Lorenz ran into Hanssen again. "I just made it," he told his friend. "He figured it out," Lorenz concluded. "Bob made exactly what he needed to get into Northwestern." Not for the first time, the underperforming Hanssen had slipped through the cracks.

4

SEARCHING FOR DIRECTION

In the fall of 1966, Bob Hanssen enrolled at the Northwestern University Dental School. It was a demanding four-year program that could not have been more different from the relatively relaxed pace at Knox College. Courses were scheduled from eight A.M. through five P.M., with no study breaks in between. The biochemistry, bacteriology, and gross anatomy courses were particularly challenging, and students were required to spend a lot of time in the preclinical lab, where they studied the morphology of teeth and worked on simulated models of the human jaw. "You were under a lot of pressure to be in the top of the class, so you could go on to your specialty," said Dr. David Neil, who graduated from the school in 1974. Another enticement for remaining enrolled was that the war in Vietnam was escalating. Every young man who had reached the age of eighteen and was not lucky enough to have a study exemption was eligible to be called up.

For a while, Hanssen showed interest in his classes. But the work was intense, and as he began his second year, his friends began to question whether he had the necessary motivation to continue. "I never had the feeling he really wanted to be a dentist," said Hoschouer. Hanssen's disinterest had a personal dimension, too. "I think what he didn't like was that that was what his father wanted him to be."

Hank Wilkins agreed. "Hanssen's parents wanted him to go to dental school, but he was not enthusiastic about the idea," Wilkins said. "The goals of his mom and dad were not his. His parents had certain expectations, but he did not accomplish them because that is what his parents wanted. *He* didn't."

His heart not in it, Hanssen decided to drop out of dental school, a difficult decision to break to his parents. Years later, on May 2, 1999, he brushed aside his reasons in a sarcastic message to his Taft Alumni Association: "I didn't like spit all that much." Then, in a burst of bravado, he added parenthetically: "though I was in the 98th national percentile of my Dental Board exam." In fact, Northwestern's dental students took their licensing board exams in two parts, the second of which came at the end of the fourth year. Hanssen may have done well on the first part of the exam, but he dropped out of dental school at the end of his second year, so his exam results no longer mattered.

While Hanssen's professional path was filled with uncertainty, his personal life had taken a much happier direction. He tried to enlist in the Navy's dental program in 1967, but he did not pass the eye test and was classified 4-F. His eyesight meant there was no risk of going to Vietnam and, much to his delight, Bonnie agreed to marry him. Their wedding took place on August 10, 1968, the ceremony held at St. Paul of the Cross, a large Roman Catholic church in Bonnie's hometown of Park Ridge, Illinois. Jack Hoschouer was Hanssen's best man. "I was the only guy on his side," he said. Tom Kozel, who also attended the ceremony, remembered a big reception afterward at the Chicago Athletic Association. The newlyweds headed off for a honeymoon in Mexico before settling into a one-bedroom apartment on Winthrop Avenue on Chicago's North Side.

In Chicago, and in the nation at large, it had been a year of enormous political upheaval. Dr. Martin Luther King, Jr., was gunned down in April, and Chicago's West Side, like many inner-cities across

America, went up in flames. On June 4, after winning the vital California Democratic primary, Bobby Kennedy was shot after making his victory speech in the Ambassador Hotel in Los Angeles. Two days later, he died. Richard Nixon, the GOP presidential candidate, promised he would end the Vietnam War within four years. But for many already disillusioned Americans, that was not soon enough. The country was furiously dividing into bitter factions, not only Democrats against Republicans, and hawks versus doves, but also parents against their college-age children.

Two weeks after Bob and Bonnie Hanssen were married, tens of thousands of young people gathered in Chicago, where the 1968 Democratic National Convention was being held, to express their growing frustration with the role of the U.S. government in the escalating war in Vietnam. Many protesters were supporters of Eugene McCarthy, whose unsuccessful campaign for the Democratic presidential nomination had galvanized the antiwar movement. Chicago's streets and parks became an arena for bloody clashes as helmeted police with riot sticks and tear gas closed in on angry demonstrators gathered in Grant Park. When the three major television networks flashed scene after scene of the violence into American living rooms, the nation recoiled. Bob Hanssen wrote a letter to Jack Hoschouer in which he said the newspapers were using the term "pigs" to describe his father's own Chicago police. The letter suggested to Hoschouer that Hanssen agreed with the epithet. If so, he put himself in the company of many people who still supported the war but deplored the tactics of the Chicago police.

Yet for Hanssen, the antiwar movement was mostly a diversion. His concerns were far more personal. He was a dental school dropout with a wife and an unsettled future. Returning to the state mental hospital was not a serious career move, but it wouldn't hurt to have a job while he figured out what to do next.

Hearing that his son was back at the hospital, Howard Hanssen was keenly disappointed. And Howard, not a subtle man, was sure to have communicated to Bob his displeasure. "One of the worst things a father can do to a son is express disappointment," said Dr. David

Charney, a psychiatrist interviewed for this book before he joined Hanssen's legal defense team. "Anger is easy for men. It is honorable, manly, hot, better than other emotions like rejection, humiliation, fear, shame. But if a son feels he has disappointed his father, first he is hurt and then he says, 'I'll show you.'"

In the fall of 1969, Bob enrolled in Northwestern University's School of Business, which offered a two-year MBA program. For Hanssen, strong in math, studying accounting was a practical choice. Northwestern's graduate program had a bevy of smart professors, including marketing guru Phil Kotler, as well as an interesting mix of students — some right out of college, and others who had worked for several years. Even for students with math and science backgrounds like Bob Hanssen, business school presented new challenges. There were difficult courses in statistics, probability, and organizational theory, as well as marketing and finance. "The work was not intellectually complex, but the workload was overwhelming, much higher than anything I had experienced before," said Ron Sanderson, a business school classmate of Hanssen's, who had majored in engineering as an undergraduate at Northwestern. "There were long hours and huge amounts of homework."

Sanderson said that when he met Hanssen, the dental school dropout struck him as "brilliant but kind of aloof," with a sarcastic sense of humor. "He often seemed to have a smirk on his face in class." Sanderson said Hanssen looked at the basic concept of a job and a career as a game, "and he was trying to figure out what the game was. He never seemed entirely happy with the whole idea of getting an MBA and working for a corporation." As the young men got to know each other better, sometimes riding together to class on the elevated train, Hanssen impressed Sanderson as being "extremely confident and self-assured," someone who always viewed himself on a different level from everyone else. "He seemed to think he was smarter than others," Sanderson said. "It could have contributed to him being, as far as I could tell, not part of any one group. He was a bit of a loner." Also, Hanssen was married, which would cause him to socialize less with his classmates.

One of Sanderson's most vivid memories of his classmate involved a discussion about dental school in which Hanssen explained to him how he had cut open a human cadaver. "Bob explained to me that what you do is look at the person and remember that no one is there, that the shell of the person is all that remains."

When Hanssen graduated in 1971, with a master's degree in accounting and information systems heading his résumé, he found a job in Chicago as a junior auditor for Touche Ross & Company, one of the nation's Big Eight accounting firms. His salary was about $10,000 a year and quickly rose to over $12,000.

For many people, a position in a firm like Touche Ross was a plum job, setting the stage for a productive business career. But while he no longer had to deal with teeth and saliva, the work failed to excite Hanssen. There was never the kind of adrenaline rush, not even occasionally, that makes a job rewarding. Nor could Hanssen shake the realization that his accounting position carried with it none of the authority that his father had with the Chicago Police Department. In his work with the Intelligence Unit, Howard Hanssen had access to countless dossiers and government documents. He was privy to the sort of confidential information that set him apart from his colleagues in the police force, affording him stature and prestige. In his own world of law enforcement, in his own bureaucracy, Howard Hanssen had power.

Howard Hanssen was due to retire from the force in July 1972. Bob's parents had long planned to move to Florida, and when the time came they sold their house on North Neva Street to Bob and Bonnie. It was the home where Bob had been raised, in the neighborhood where he had grown up. Many of his classmates had moved away, but the surroundings were comfortable and familiar. Bonnie's family still lived in Park Ridge, not a ten-minute drive away. It all made good sense. To help pay the mortgage, Bonnie got a job as a clerk in a social services agency run by the city. Her uncle, a local judge, helped her find it. Bob and Bonnie also asked Bonnie's sister, Jeanne, and her Greek-born husband, George Beglis, to move in with them. Space was tight, so the two young couples did some modest remodeling. They had a wall removed, shifting the stairs from the back room to the front room, which gave them more living space. The Hanssens would soon need more. Bonnie had suffered a number of miscarriages earlier in their marriage, but in July 1972, Bonnie gave birth to their first child, Jane.

It's not at all unusual for a young person to follow in a parent's footsteps. But it is a bit surprising when the child has spent much of his life at odds with the parent. Nevertheless, Bob Hanssen figured one way to trump his father might just be to join him, to chuck aside everything he had done and become a cop. But the idea wasn't just to become a cop, but to take on Howard on his home turf. Now the rules for comparison would certainly be clear. Now it would be impossible not to acknowledge what Bob knew to be perfectly obvious but which others didn't seem to grasp: that the son could eclipse the father.

Three months after his father retired, Bob Hanssen signed up for the force. At twenty-eight years of age, a college graduate with a graduate degree in accounting, he was way overeducated to be a Chicago beat cop. Nor would breaking up street brawls and writing traffic tickets give him the same rush that it did most other cops. His father's intelligence beat, on the other hand, was a field that Bob Hanssen understood, a place where he felt comfortable and could excel. It was also a natural outlet for his geeky side, as it required considerable technical skills.

By any measure, the social changes of the early 1970s were unsettling to conservative families, even young ones like the Hanssens. The antiwar movement had given rise to a growing sense of dissatisfaction on many fronts. The American workplace was becoming a confusing battleground for the sexes. Most men and many women had little understanding of the profound effect the women's movement would have on every aspect of American life over the next three decades. High inflation, skyrocketing interest rates, and stagnant growth forced millions of couples who had planned to live on a single paycheck to accept that the only way to maintain their standard of living was for both spouses to go to work. With baby Jane, the Hanssens' finances were strained. Rookie pay was $877 a month, considerably less than the $1,050 Hanssen had been making as an accountant.

In October, Bob Hanssen formally enrolled in the Chicago Police Training Academy, located in a decrepit building on West O'Brien Street that once served as a hospital for Civil War soldiers. Recruits spent twelve weeks getting in physical shape to become officers. They also took classes in law, police procedure, marksmanship, and writing reports. When classes were over, recruits were put on probation for a year and usually assigned to a field training officer. Occasionally, the city drew on academy recruits when they needed extra help at a specific location or on the streets.

It was an interesting moment to be a rookie Chicago cop. There were at this time no less than three federal grand juries investigating scandal and corruption in three separate Chicago police districts. Mayor Richard J. Daley, who had basically run a one-party dictatorship for years, saw his political power waning. In an attempt to boost his popularity with disgusted voters, Daley asked one of his trusted aides, a consultant named Jack Clarke, to put together a team of undercover officers to ferret out corruption in the department. It was called the C-5 Unit and was headed by Mitchell Ware, a controversial deputy superintendent. Because he needed officers who would not be recognized on the street, Clarke looked for recruits in the police academy, rookies whom he could train in sophisticated undercover techniques.

When approached, Bob Hanssen volunteered gladly.

"I did some research and found his father was in the police department assigned to intelligence," Clarke recalled. The mayor's confidant was also impressed by Hanssen's manner. "He was quiet, a slick kind of guy," Clarke said. "He was analyzing me analyzing him. He was extremely confident. His college education showed through. We had asked for some Fords and they sent us a Mercedes."

Clarke had a story he liked to tell his young charges about the kind of person who constitutes a good undercover agent. It was about an Irish boy with a dark complexion who learned to speak good Spanish.

Clarke hired the kid to infiltrate a Mexican gang on the West Coast and gave him a pseudonym: Ramon. "I would explain that because of his language ability and adaptability, he did a great job undercover," Clarke said.

Another young officer who worked with the C-5s was Ernie Rizzo. Rizzo said he met Hanssen when they attended a secret electronics surveillance school that operated in a dusty Chicago storefront disguised as a television repair shop. "He was brilliant, far too smart to be writing traffic tickets on the street," said Rizzo, now a private investigator. "You could tell this by the way the guy carried himself." Rizzo said that their teacher was a Japanese intelligence specialist who taught them his cardinal rule of spying: "Never let the left hand know what the right hand is doing." The instructor also taught his students that if they were doing surveillance work and were stopped by a member of another law enforcement agency, they should be careful about what they said. "If you tell him you are doing an insurance investigation, it turns off most law enforcement guys," Rizzo said. "If you are ever stopped doing one of these black-bag jobs, tell him this and chances are he will forget about you."

Rizzo remembered Hanssen as a rigid man, "the last guy you would ever want to team up with. Sometimes you want to rap with one of the guys, but not him." Hanssen also had a reputation as a Commie-basher: "He was always going around saying the Russians are rotten bastards."

Not surprisingly, the C-5 Unit was despised by other officers, who saw the work of spying on fellow cops as the essence of disloyalty, a betrayal of the badge they wore so proudly. The C-5s weren't helped by their inexperience, but even veteran officers would have found the assignment difficult. "They were the gang that couldn't shoot straight," said Richard Sandberg, a retired Chicago police lieutenant who spent thirty-three years with the department. "The C-5s were a disaster." Most Chicago cops did not consider C-5s competent enough to hack it on the street. Bob Hanssen was, once again, an outsider.

Over the months, Jack Clarke, who was initially impressed with Hanssen's skills, became increasingly suspicious of his motives. "He was cunning," Clarke said. "He would come to me with a technical question and turn the conversation to politics. He'd ask a question about Mayor Daley or the Democratic Party. He never said anything about himself. He roped people into conversations about themselves." Eventually, Clarke began to suspect that Hanssen was a plant, that he might be working for the police department brass who wanted to know who was being investigated. Or maybe even for the FBI. "I knew the FBI was coming down on me," said Clarke, who later was convicted and imprisoned for tax evasion. (It was a sign of how corrupt the Chicago police were that the man in charge of the C-5 Unit was crooked.) "So I decided to send the FBI a message. I told him . . . to go down to the federal building and apply for a job."

Hanssen did just that. And why not? After so long in the wilderness, he had found his calling. His first training in the field of law enforcement had been a course on how to spy on fellow cops, how to deceive those who think you are one of their own. As a cop, Robert Hanssen had tasted what it was like to be a double agent. Perhaps not surprisingly, it felt quite natural. After all, hadn't he always been a bit undercover? Pretending, exaggerating, hiding, stretching the truth — Hanssen had been doing this for a long time. In the past, people had ignored him without consequences. But now he wouldn't just be in the background — he'd be in the background *taking notes*. And then, later, they'd never see it coming when he pinned them to the wall. Yes, he liked this secret world, this intriguing and ever more intricate veil of shadows. Bob Hanssen was hooked.

5

JOINING THE FBI

Ask any present or former FBI agent of a certain age why he — it is almost always a man — joined the FBI, and if he grew up in the forties or fifties, he'll tell you how he was transfixed by the radio melodrama *The FBI in Peace and War*. If he was a kid in the 1960s and early seventies, what he knew about the FBI came from Efrem Zimbalist, Jr.'s, portrayal of the dauntless Inspector Erskine on the long-running network TV series *The FBI*. Generations of young Americans knew that although they couldn't hope to be knighted by King Arthur or deputized by Marshal Dillon, they could hope to carry the simple round gold badge that proclaimed "Fidelity Bravery Integrity."

Bob Hanssen shared those fantasies. "He had a very idealistic view of the FBI," said his friend Jack Hoschouer. In late 1975, following Jack Clarke's advice, Hanssen made up his mind to apply for a job with the bureau. His father strongly disapproved, not wanting his son to start all over again just to be another kind of cop. Also by then, the bureau's image, which J. Edgar Hoover had masterfully burnished over the course of his forty-eight-year tenure, had been battered beyond recognition. Once hailed as a unique government institution with a long, proud tradition, the bureau was now a pariah, assailed daily by politicians and the press as a rogue agency that had hounded hundreds of thousands of Americans, whose only offense was to criticize the conduct of the Vietnam War or to seek racial or sexual equality.

Hoover had died on May 2, 1972, and Clarence Kelley, a former FBI agent and Kansas City police chief, was confirmed as FBI director on June 27, 1973.* He was determined to redeem the bureau's reputation. Kelley made the admirable but painful commitment, for the FBI and for himself, to open the files of the Domestic Intelligence Division. These chronicled decades of FBI actions, in violation of the First Amendment guarantee of freedom of speech and association, against largely peaceful political dissenters. In addition, a breathtaking litany of FBI abuses conducted under the code name COINTELPRO (for counterintelligence program) was made public by a special Senate panel, chaired by Senator Frank Church, a Democrat from Idaho. COINTELPRO had been launched in 1956 to neutralize the activities of the Communist Party of the United States of America (CPUSA) and had expanded over the years to encompass surveillance of and covert action against nearly every activist group in the nation, from the Reverend Martin Luther King, Jr.'s, civil rights crusade to antiwar protesters and environmental activists. Conducted with minimal supervision from the Justice Department or the White House, the FBI's COINTELPRO eventually developed over 500,000 files on individuals and groups.

Most rank-and-file agents worked for the criminal investigative division, where they handled the bread-and-butter cases: white-collar crime, violent crime, organized crime, bank robberies. They had nothing to do with COINTELPRO. Most didn't even know about its existence until they read the newspapers, and then they were as

---

*Nixon's first choice to succeed Hoover, L. Patrick Gray, had been forced to withdraw his name after acknowledging that, while acting director, he had destroyed documents found in the possession of Watergate burglar Howard Hunt at the insistence of White House counsel John Dean. Gray had also allowed Dean to sit in on FBI interviews with White House aides.

angry as anybody else. Still, they suffered for Hoover's sins, because outsiders made no distinction between factions within the FBI, and the daily pounding made agents all over the country cringe whenever they opened a newspaper or turned on the network news.

It was into this turbulent world that Hanssen plunged when he was sworn in on January 12, 1976. He was exactly the kind of recruit the FBI wanted. The bureau gave preference to applicants with advanced degrees in law or accounting or at least two years of professional work experience. Most important, Hanssen meshed with the bureau's conservative, middlebrow culture. The FBI recruited working-class kids from Middle America, often Catholics, as well as people who had not been part of the sixties cultural revolution and actively detested everything about the leftist student movements. Ivy Leaguers seldom applied, nor were they courted. Hanssen fit the bill in every respect.

Hanssen attended new agents' school at the FBI Academy in Quantico, Virginia, where he struggled through the fitness course. "He had never done a lot of physical stuff, climbing ropes and doing push-ups," recalled Hoschouer. "He was very proud he had made it through."

His first assignment was Gary, Indiana, located just over the state border from Illinois on the southern shore of Lake Michigan. With its steel mills closing and much of the professional class fleeing to the suburbs, Gary was a dreary and difficult place to live. Chicago was only about thirty miles away, but rush hour could make it a sixty-to-ninety-minute drive, so Bob and Bonnie moved the family to a little house in Hobart, Indiana, a small town about ten miles from Gary. It was the first of five moves they would make together.

The FBI had an active office in Gary, where Hanssen was assigned to the white-collar crime squad. Younger agents generally concentrated on cases involving political corruption, interstate transportation of stolen property, and prostitutes who crossed state lines. There was an interesting international component to the job as well. When Soviet ships came down the St. Lawrence Seaway and into Indiana's international port to drop off cargo — steel coil, steel slabs, soy beans, and corn — FBI agents would find an excuse to board the vessels to see what else they could learn about the ships' mission. With a civilian electronics engineer, Hanssen would walk through a ship's radio room to check out the length and shape of the various antennae, using his ham radio experience to assess their frequencies and capabilities. "This is how they would determine if it was a real merchant ship carrying grain or a spy mission," explained a military friend. The position was challenging, and Hanssen, well-trained and confident, quickly became cocky and comfortable. He seemed comfortable as well in his personal life. Bonnie had given birth to two more children, Susan, a daughter, who was born in February 1974, and a son, John Christian, born in April 1977. They called him Jack.

Hanssen was changing spiritually as well. In 1976, while on an Army exercise in England, Jack Hoschouer had received a letter from his old high school buddy. Like many friends whose careers and families take them in different directions, the two men had drifted out of touch. They had not seen each other for two and a half years, but there was still a strong bond between them. Hoschouer was eager to know how Bob was faring in his new life. Was he still the same old guy who liked tinkering with electronics and cars, or had the responsibility of marriage, children, and a job changed him? One day Jack received a letter from Bob. More than twenty-five years later, Hoschouer remembered its dramatic tone and exactly how it began: "Slowly, over a long period of time, I've come to know the truth."

Hanssen had converted to Catholicism. The news, said Hoschouer, came as "a complete surprise." The Bob Hanssen he knew had been raised a Lutheran, yet he had never struck Hoschouer as particularly religious. He may even have been antireligious. Bonnie, of course, was Catholic. And as Hoschouer thought about his friend's decision, it began to make more sense: "I think all his life he had been searching for something, for some hard, well-thought-out intellectual truth. Here was a firm anchor with clear rules that ultimately point in a direction that is good." One aspect of the decision was pure Bob, said Hoschouer: "He didn't share it with me until the deed was done."

*   *   *

Within two years of converting to Catholicism, again following Bonnie's lead, Hanssen joined Opus Dei. Hanssen later told a friend that one of those who had influenced him the most about joining the ultraconservative and staunchly anticommunist Catholic society was a fellow law enforcement officer, a roommate at the Police Academy.

Although headed by a bishop in Rome, Opus Dei is essentially a lay organization. It was founded in 1928 by a Spanish priest named José María Escrivá de Balaguer y Albás, a greatly revered Church figure, who believed in encouraging people to aspire to sanctity without changing their occupation.

Opus Dei is so small that even many Catholics have never heard of it. In 2001, it has 84,000 members worldwide, some 3,000 of whom were in the United States. The organization has sixty centers, or residences, for members in nineteen U.S. cities, up from eight cities in 1975. Many centers are located in close proximity to college campuses, often Ivy League schools, where Opus Dei concentrates on recruiting new members (the smarter and the more elite, the better.) Each center holds ten to fifteen unmarried members who live in separate houses, segregated by sex. Opus Dei also runs five high schools in the United States, including two that operate in the suburbs of Washington, D.C.: the Heights, and Oakcrest. It was to these schools that the Hanssens would eventually send their children.

When Hanssen joined Opus Dei, he became a "supernumerary," one of a dizzying array of Opus Dei membership categories. There are priests, who make up 2 percent of the organization; "supernumeraries," married individuals like Hanssen who compose 70 percent; and "numeraries," single men and women who pledge to remain celibate and live in sex-segregated Opus Dei residential centers, and who comprise 25 percent. There are also "associates," often teenagers who live with their parents, and "cooperators," nonmembers who have not received the "divine calling" but contribute financially to the cause. "The Opus Dei people I know tend to be well-balanced," said Robert Royal, a conservative Catholic writer (not a member of Opus Dei) who is president of the Faith and Reason Institute in Washington, D.C. "For many people, it is a way to have deep religious commitment and still operate in the real world."

The stated purpose of Opus Dei, whose Latin name translates as "the work of God," is to help people find holiness in their daily life, regardless of career, social status, or occupation. Its promise, like the reflected hues of a rainbow, is that those who strive for perfection in all aspects of their life will come closer to finding God. "It is a demanding life of devotion, piety, prayer, and commitment to work well," said the Reverend C. John McCloskey III, a charming, outgoing Opus Dei priest in Washington who once worked at Princeton University. "Opus Dei is a vocation, a divine calling from God. It's hard to join but easy to leave." Dr. James Egan, a respected child psychiatrist, said the most significant aspect of the organization is the effort members like himself must make in their daily life. "We are all called to try to be holy and saintly," he said. (Translation: this is an organization that demands plenty of time.)*

Opus Dei members consider themselves recipients of a personal invitation from God to serve the church and sanctify themselves by following a particular, and rather unusual, path. "Imagine a woman who receives in the mail an invitation to a party to which all of the guests are asked to bring a dish," wrote John F. Coverdale, a law professor at Seton Hall University School of Law, author of an Opus Dei–published booklet titled "On the Vocation to Opus Dei." "A few days later she receives a personal visit from the host pressing her to come to the party and asking her to bring the seafood salad she makes especially well. . . . Similarly, in recognizing a vocation to Opus Dei, a person becomes aware that Jesus addresses to him or to her personally his invitation."

Opus Dei is considerably more complicated than seafood salad. Its philosophy is rigidly doctrinaire, a tightly prescribed framework of

---

*Egan and his wife, Anne, knew Bonnie and Bob Hanssen casually from various Opus Dei functions. "In my conversations with him, he often talked about the threat of communism," Egan said. "He was vocally anticommunist."

restrictions and platitudes that can prove a rather tricky terrain, particularly for those who may be vulnerable and uncertain of their own personal direction. It is a virtual spiritual garden, unburdened by thunderstorms or clouds of doubt. As a result, many — though not all — members radiate a sense that they are more Catholic than Catholic, which other Catholics deeply resent. "Bob Hanssen is not like any Catholic I ever knew," said a Washington, D.C., resident who was raised in the faith. "He tried to be more Catholic than the Seans, O'Malleys, and Finnegans." Bonnie Hanssen could be the same way. At times, she even seemed to vie with other Opus Dei women over who sat closer to God. Some Opus Dei women resented this.

James Martin, an associate editor of the Catholic magazine *America,* called Opus Dei "the most controversial group in the United States today." And even Opus Dei members concede that they are often regarded with fear or suspicion. "You get on the defensive about Opus Dei," said Oakcrest's Barbara Falk. Why? "Because we are a big threat. We are in-your-face Catholics. We really want people to be better."

Opus Dei members adhere to a strict "Plan of Life." Brian Finnerty, the organization's national spokesman, described this as being "the set of priorities that people sprinkle throughout their day." These include attending daily mass, saying the rosary, reading from the Gospels, and reciting the Angelus, a traditional prayer said at noon in honor of Mary, the Blessed Mother. Opus Dei members are expected to attend a weekly "Circle," a class devoted to discussions of such aspects of Christian living as prayer, conscience, and the importance of order in one's life. An Opus Dei priest is present for anyone who has not yet gone to confession.

Opus Dei members meet privately every two weeks with a spiritual director. They also must attend monthly "Evenings of Reflections," mini-retreats of several hours devoted to making spiritual resolutions for the coming month. The purpose of these meetings, said Finnerty, is "to recharge one's religious batteries." Bonnie Hanssen often led evening seminars for women at her family's Catholic church in Virginia. Members attend an annual retreat, often over a weekend.

Every year on March 19, the Feast of Saint Joseph "the Worker," members renew their vows to Opus Dei. "Work is central to our lives," Dr. Egan said.

Another aspect of daily life for Opus Dei members is corporal mortification, which requires denying themselves certain comforts and inflicting varying degrees of physical pain on their bodies. Finnerty said that mortification is primarily concerned with such small daily acts of thanksgiving as getting up when the alarm clock rings or keeping your desk in order. "Life rotates around trying to do your work well, setting aside time for prayer, deepening friendships, and bringing yourself closer to God," he said. Many members also wear a device around their upper thigh called a "cilice," a spiked chain that bites into their flesh, leaving its mark. They wear the cilice for two hours every day but Sunday to remind themselves of their sins, as well as those of others. "These practices of Christian asceticism are no more harmful to health than are athletic training or the diets followed by many to improve their health and appearance," wrote John Coverdale. "They are a way of sharing voluntarily, in a small way, in the suffering of Jesus Christ." Corporal mortification, he wrote, "is compatible with a cheerful, contented life."

Members also flog themselves weekly on the back or buttocks with a "discipline," a whip made of woven cord. Female numeraries sleep on boards placed on top of a mattress because Founder Escrivá believed that women have emotions that are difficult to tame. Male numeraries sleep once a week on the floor. "Some people do corporal mortification because it helps put the soul in charge of the body instead of the other way around," said member Irene Dorgan. "Most religious people find that some corporal discipline helps them achieve a spiritual high."

Dorgan has been a numerary since the late 1970s. An outgoing woman with a friendly personality, she lives in an Opus Dei house for women in New Rochelle, New York, and works in an Opus Dei–founded center for girls called Rosedale, located in the Bronx. "We help people with spiritual mentoring," said Dorgan, who, like all numeraries, contributes her entire salary, less a portion for personal expenses, to Opus Dei. "Our sole purpose is to show that holiness is possible to achieve in any profession or line of work. It is very idealistic but also realistic."

Another aspect of Opus Dei tradition is that female numeraries cook and clean for their male counterparts, who must vacate their residence when women enter. Irene Dorgan said that women are trained for these jobs, often studying for degrees in hospitality from Opus Dei schools in Boston and Chicago. An Opus Dei official, who was perfectly at ease in discussing most aspects of the organization, got tongue-tied and insisted on anonymity when it came to explaining why only women are expected to clean and cook. "Part of the idea is to provide a homelike atmosphere in the center, not the feel of a hunting lodge," he said. "Look," he added, "the number of parish priests who vacuum is not that high. Seeing that as a task can lead people closer to God."

Opus Dei may not have high standing with feminists, but it does with the Vatican. In 1982, Pope John Paul II granted the organization the status of "personal prelature," a term for an ecclesiastical jurisdiction that covers all persons in Opus Dei rather than a particular region, such as a diocese. It is the only personal prelature in the Church. In 1992, the Vatican showed favor to them again by beatifying Escrivá in a ceremony in St. Peter's Square. Because this major step toward sainthood came only seventeen years after his death, in 1975, and because such an enormously respected figure as Pope John XXIII had not received such honor, the beatification was immediately controversial.

Opus Dei is sometimes criticized for putting a priority on recruiting affluent people, for employing tactics that are overly aggressive, and for exerting excessive control over young members. "We don't call it a cult but it's cultlike," said Dianne diNicola of Pittsfield, Massachusetts, executive director of an organization called Opus Dei Awareness Network, which she started ten years ago to educate families of young people recruited into their ranks. "They control people. They're not truthful."

Ann Schweninger, thirty, of Columbus, Ohio, belonged to Opus Dei for six years before she dropped out in 1992. When she joined the organization in 1986 at the age of sixteen, Ann was a student at Oakcrest. For two years Ann carpooled to school with Jane and Sue Hanssen, the two oldest daughters of Bonnie and Bob. Ann said Opus Dei was very aggressive in recruiting her and Sue Hanssen before they became Opus Dei members. (Sue, still a member, wears a cilice and sleeps on a board.)

Schweninger said that when she joined Opus Dei, she had not been told that she would have to wear a cilice on her thigh or flog herself with a whip. She also did not know that the director of the Opus Dei center where she lived would read her mail, incoming and outgoing. "There was a lot I didn't know," she said. "They did it for control. They wanted to know your concerns, so they would know what was happening in your life." She said that when she dropped out, the director of her center wrote her repeatedly, asking her to return. She did not.

Bob Hanssen would have had a much better idea of what he was getting into when he joined Opus Dei, because he had been exposed to the strict regimen and expectations through Bonnie's family. The organization brought structure to lives that had zigged and zagged in too many directions. It's also logical to believe that this small, elite society gave him an identity that he craved, a larger group of friends than he had ever had, the social acceptance he had never known. It may have also reinforced his holier-than-thou image of himself. And of course, there was the whole conspiratorial nature of Opus Dei — a *very* private army, indistinguishable from everyone else but full of secrets, rituals, and providing the sort of confidence that comes from knowing who shall be drowned and who shall be saved.

At the age of thirty-six, Hanssen was finally on his way.

6

NEW YORK FIELD

Almost three years as an FBI agent in Gary had pumped up Hanssen's sense of his own destiny. But his real Welcome-to-the-NFL moment came on August 2, 1978, when he was transferred from Gary to the FBI's New York field office (known in bureau parlance as "NYO"). The move was a painful one. There was no cost-of-living adjustment, and salaries of younger agents were on a par with those of Manhattan sanitation workers. Single agents lived in boarding-houses in Brooklyn or shared apartments in the more unfashionable neighborhoods of Manhattan. Those who were married with children lived an hour or two out of town, in the far reaches of West-chester County, Monmouth County, New Jersey, or even northern Pennsylvania. Some agents were forced to apply for food stamps, jump to the better-paying New York Police Department, or go back home and practice law. But New York was the big time, for FBI agents as for aspiring artists, actors, writers, heroin dealers, con men, and hookers. Bob Hanssen, who only a few years earlier had been an accountant going nowhere, was now at the center of the action, the Carnegie Hall of G-man-dom.

Hanssen was assigned to a white-collar crime squad, where he met Tom Pickard, also an accountant by training, who was about to go undercover as a crooked bookkeeper in the groundbreaking ABSCAM sting. Pickard thought Hanssen "a good accountant," who kept his desk very clean, not an insignificant thing in an organization that made a sacrament of order. And while Pickard didn't find Hanssen particularly personable, he thought Bob had promise.

Hanssen settled in, finding a modest house in Scarsdale, a toney town twenty miles from Manhattan. Bonnie would soon be pregnant with a fourth child, adding to the financial crunch. Bob may have tried to calm Bonnie by explaining that if he were ever to climb the ladder in the bureau, NYO was the place to be. But if she visited her husband's new workplace, she might have had her doubts. From the outside, the old Lincoln warehouse building at 201 East Sixty-ninth Street wasn't much to look at. Every day, scores of people walked past its sooty façade without realizing that FBI's flagship office loomed high above, over a shoe store, an administrative office of Hunter College and a branch of the New York Telephone Company.

The interior wasn't any better. The inside walls were two shades of institutional green. The floors were gummy green asphalt tiles over concrete, heavily reinforced decades before to accommodate the suits of armor, mahogany armoires, marble statuary, and mountains of whatnots accumulated by the previous tenant, publisher William Randolph Hearst, during his European travels. Some eight hundred special agents, about a tenth of the bureau's force, worked in the office. The street agents, also known as "brick agents," were relegated to windowless bullpens furnished with battered metal desks, decrepit metal swivel chairs and a few, heavy, black dial telephones with long cords, so they could be passed from desk to desk. The ceiling lights sometimes dripped a brown acid, thick as motor oil, onto the desktops. The phones were strictly for local calls. An agent needing to call, say, Newark, had to go to the supervisor's cage, dial the FBI operator, give her a case number and request a long-distance line. If he went to the men's room and missed the operator's call, he went back to the bottom of the list. On busy days, you might see quite a few guys standing around with their legs crossed. If an agent had to *go* to Newark, he lined up for one of the two or three cars earmarked for his squad, which usually numbered eighteen agents. These were parked in a garage a good fifteen blocks from the warehouse, and on raw winter days even the toughest street agents were tempted to stay inside and seek permission to phone.

There were a few rudimentary computers on the market when Hanssen showed up, but people joked that the bureau's real motto was "yesterday's technology tomorrow." Agents took notes by hand, then dictated them to a stenographer, who made multiple copies of all reports and 302s (summaries of interviews related to particular investigations) with thick packets of carbons and flimsies. The master copy went into a World War II–vintage file cabinet, to be preserved till the end of recorded time. Carbons went to headquarters by mail or, if high priority, "airtel" — airmail. An extremely time-sensitive message went out by Teletype.

Shortly after Hanssen's transfer, the office was filled, literally, by Neil Welch, a brusque, lantern-jawed, refrigerator-size veteran who introduced himself to the troops by announcing he was going to rebuild the office "brick by brick." As the assistant director in charge (or ADIC — usually pronounced "A-dick") Welch enjoyed the meager perks the FBI had to offer: a big, tenth-floor office with windows, a desk of real wood, a padded swivel chair, and his own government cars. But even Welch dared not occupy the grandest suite, which featured a desk the size of a Rolls Royce and the floor's best view of the Manhattan skyline. This was Hoover's office, which the late director had used when he stopped by Manhattan on his way to the racetrack. Though Hoover had been dead six years, his office was maintained and his ancient Cadillac limousine garaged as though he might return at any moment.

Colored shirts and sideburns, banned by Hoover, now blossomed here and there, and some people actually used blue ink, a color that had been reserved for the exclusive use of the director. The rule forbidding coffee at desks stuck. It didn't make a lot of sense; what could a little brown spill do to those ugly, indestructible desktops? How many hours were wasted by agents trudging to Kasey's Kitchen on East Sixty-seventh for a caffeine fix? But that's the way it had always been, so some of the more hidebound supervisors saw no need to change it. And in fact, many agents, especially the more senior ones, took a perverse pride in their Spartan working conditions. Like the Marine Corps (whose veterans were amply represented in the post–World War II bureau), it was an article of faith at NYO that a few good men and women could beat the devil himself, if need be fashioning weapons out of penknives, chair legs, and shoestrings.

There was some truth to this conceit. The agents of NYO faced the worst of everything — the most audacious, most rapacious, most cunning, and just plain most of every brand of criminal, from Wall Street's high-rolling predators, to the five La Cosa Nostra families that reigned over organized crime internationally, to Puerto Rican terrorists, the Black Liberation Army, and the wacko hostage-takers featured in the film *Dog Day Afternoon*.

Though outnumbered on every flank, NYO's aggressive criminal squads counterattacked ferociously and resourcefully, especially after Clarence Kelley's "Quality over Quantity" initiative liberated them from the trivial cases so beloved by Hoover because they boosted his arrest statistics. By the mid-1970s, NYO agents were employing controversial but devastatingly effective tactics on every front. And ambitious young agents knew that NYO was where reputations were made, and stars were born.

For instance, in 1975, the organized crime squads launched a sweeping investigation of corruption and racketeering on the East Coast waterfront. Kelley had lifted Hoover's prohibition on undercover operations, so a fuzz-cheeked agent named Louis Freeh went undercover — in fact, clad only in a towel — to frequent a steam bath favored by Michael "Big Mike" Clemente, a Genovese family crime boss suspected of extorting waterfront businesses. The sweeping UNIRAC (for "union racketeering") undercover case went on for six years, in the end securing the convictions of 125 mafiosi, businessmen, and officials of the International Longshoremen's union locals from Boston to Miami. For years to come, UNIRAC would serve as a model for federal probes into organized crime networks. And, of course, Louis Freeh would eventually find himself in grander offices.

For all the adrenaline pumping through the Lincoln warehouse, a lingering gloom hung over the place. There's a dark joke in law enforcement: "Big cases — big problems. Little cases — little problems. No cases — no problems." Because NYO agents had handled most of the big, politically sensitive cases in the sixties and early seventies, they found themselves targets of the post-Watergate investigations. Agents who brown-bagged their lunches because they were too broke to go to the deli were now forced to hire lawyers, answer to internal investigators, and go before a federal grand jury.

In the spring of 1976, FBI headquarters personnel reviewing files came upon a cache of documents that showed that an NYO intelligence unit known as Squad 47 had engaged in break-ins and other illegal acts during the search for Weather Underground members who had disappeared in 1969 after taking credit for bombings at the U.S. Capitol, the Pentagon, and other federal facilities.

The newly discovered documents directly contradicted denials by director Kelley and other high FBI officials. In April 1977, soon after the Carter administration came in, John J. Kearney, the former supervisor of Squad 47, was indicted for some of the warrantless intrusions. Though Kearney had been retired for four years, his indictment sent shock waves through NYO. Unlike many of the targets of COINTELPRO, the Weathermen fugitives were not peaceful protesters but admitted bombers. Further, as the FBI agents saw it, Kearney and his agents had been acting on orders from headquarters, which in turn was acting on orders from the Nixon White House to stop the bombings.

As Kearney arrived to be arraigned on April 14, 1977, some three hundred FBI agents took to the streets, staging an unprecedented demonstration outside the U.S. Court House in lower Manhattan. Kelley issued a statement praising Kearney and asked Griffin Bell, President Carter's attorney general, not to seek charges against Kearney's subordinates in Squad 47. Bell agonized for some time before taking

Kelley's plea to heart and killing the planned indictments of lower-level agents. But they could still be disciplined or sacked.

William Webster, a federal appeals court judge from St. Louis who was sworn in to succeed Kelley on February 23, 1978, inherited the unwelcome task of leading an internal investigation into about seventy NYO field agents and supervisors who knew of or took part in Squad 47 activities. In December 1978, Webster fired two FBI supervisors, demoted another, suspended a fourth, and censured two street agents. Fifty-eight street agents were cleared.

Until John Kearney was indicted, the notion that an FBI agent could be charged with crime for following orders had been unthinkable. Generations of agents had labored under an unspoken agreement that if they put up with low government salaries, long hours, lousy working conditions, and danger, the government would take care of most of their needs, so long as they did nothing for personal gain. With the Kearney indictment, they lost their innocence. The Justice Department came to be viewed as an adversary, a capricious, overzealous, and dangerous one that shifted with the prevailing political winds. Agents and prosecutors had always argued with each other, but after 1976, many agents suspected that department attorneys would not only mangle their best work but sell them out in a heartbeat. The FBI that Bob Hanssen had joined was not a happy place.

Hanssen scarcely had time to find his way around his white-collar crime squad division when, in March 1979, he received a transfer to the Soviet intelligence division, also known as Division Three.

This was not a tribute to his command of Russian or anything else. Division Three and its brother, Division Four, which covered Eastern Europe, China, Cuba, and Vietnam, were going through a period of severe reduction. The oldest hands were being forced out by a recent law setting fifty-five as the mandatory retirement age for FBI agents. (This was Congress's way of heading off another septuagenarian autocrat like Hoover.) Also, as New York housing prices skyrocketed, mid-career agents in their thirties and forties were demanding transfers to less glamorous but more affordable cities. One agent working there at the time said of his colleagues, "Getting a job in the FBI was a childhood dream. But they were miserable, their wives were miserable, and the dream wasn't working."

The bureau drafted the youngest, poorest paid, and least experienced agents to fill the empty desks. Hanssen was one of the few rookies who actually wanted to be there. Most others rebelled at the excruciatingly slow pace of the work — the criminal-division agents called the foreign counterintelligence (FCI) components Sleepy Hollow — and spent their waking hours plotting their return to chasing "real" bad guys. They had joined the FBI to handcuff mob hit men and international con men, not to tail Soviet intelligence officers playing hooky to browse the consumer electronics stores near Times Square.

But not everyone felt that way. Career FCI agents, while acknowledging that there were periods of tedium that could test a saint's patience, felt they were answering their nation's call, as their fathers and older brothers had done when they enlisted to fight Hitler and Hirohito. "We were on the front line of the Cold War," said R. Jean Gray, special agent in charge of Division Three from 1980 to 1984. "They were true believers," said Jim Burnett, who worked in NYO intelligence in the early eighties. "They were the good guys, going after the bad guys on a holy mission."

Hanssen ingratiated himself with his supervisors with his somber patriotism, though he made a few of his colleagues a little uneasy by adorning his desk with a crucifix and letting it be known, said Tom Burns, who worked with Hanssen on and off from 1978 until almost the moment of his arrest, "that he believed communism was the incarnation of Satan in the world."

Still, Hanssen was a cerebral type, who seemed as fascinated by the intricacies of FCI work as they were. Agents who gravitated to the intelligence division didn't care how many arrests they made. What fed their adrenaline habit was matching wits with the other superpower's best and brightest. If you could play chess with Bobby Fischer, why would you waste your time in a pinball arcade? Or, as Jean Gray shot back when his counterparts on the criminal side boasted about their scores against the Mafia, "Yeah, well, you're playing Podunk U., and we're playing Notre Dame." Agents in FCI held themselves out as the bureau's intellectual elite. They read the *New York Times* and kept up on world events, while scoffing that if the knuckle-draggers in the criminal squads read anything at all, it was the *New York Daily News,* and the sports pages at that. They gossiped that FCI agents routinely scored so much higher on IQ tests than their criminal-division counterparts that the FBI Academy stopped giving the exam. Hanssen seemed a good fit. He was often seen in the intelligence division's small library devouring spy books, Marxist-Leninist tomes, classified analyses and reports produced by the intelligence community and the White House, and the richly detailed overnight surveillance logs.

If Hanssen hoped to make a name for himself in the spy game, he landed in the right place at the right time. New York was the overseas hub of operations of the KGB (Komitet Gosudarstvenoy Bezopasnosti, or Committee for State Security) and the GRU (Glavnoye Razvedyvatelnoye Upravlenie, or Chief Intelligence Directorate of the General Staff). There had never been a larger or more aggressive concentration of Soviet spies in the city than in the late seventies and early eighties. "We were going full bore," said Ed Curran, a supervisor in NYO Division Three. The reason was the presence of the United Nations headquarters, which offered the Soviets unparalleled access not only to Americans but also diplomats from all over the world. Some of these would eventually rise to the top ranks of their governments or to positions where they could get their hands on important political, military, or technological data.

Inside the UN complex, KGB officers, operating under diplomatic cover, enjoyed free rein, because UN rules banned FBI and CIA personnel from the premises. The Soviets exploited that edge to the fullest. In 1979, former UN undersecretary Arkady N. Shevchenko, the highest ranking Soviet diplomat ever to defect to the West, confirmed that as many as 300 KGB officers were stationed in New York, among them Viktor M. Lesiovsky, special assistant to then UN Secretary General Kurt Waldheim. The FBI estimated that a third to a half of the Soviet diplomats assigned to the Soviet Mission to the UN were trained KGB and GRU officers. "The Soviets were like sharks in an aquarium," said Jean Gray. The KGB planted hundreds more intelligence officers inside the UN Secretariat, using the cover of international civil servants, New York–based U.S.S.R. commercial fronts such as Amtorg, the Kama River Purchasing Commission, Aeroflot, Sovfracht freight forwarding company, the Soviet travel agency Intourist, the Soviet news service Tass, and visiting cultural, scientific, and trade missions. Most worrisome to the FBI were the unknown number of "illegals," intelligence officers without official cover, among the tens of thousands of Russian nationals who had recently immigrated to the United States, thanks to the Nixon administration's détente with the Soviet Union, the Helsinki Accords, and liberalized U.S. immigration laws.

The FBI operated around the edges of the UN, the Soviet Mission at 136 Sixty-seventh Street between Third and Lexington avenues, the commercial establishments, the Soviets' weekend retreat (a once elegant mansion in Oyster Bay, Long Island), and the Russian immigrant community. Lower-level Soviet diplomats lived dormitory-style in the Sovplex, a walled residential compound in the Riverdale section of the Bronx; but many higher-ranking diplomats and commercial officers, including senior KGB people, used their clout to keep their Manhattan apartments. The FBI posted stationary surveillance teams — mostly support personnel, with few agents supervising them — "in the buckets," meaning in rented spaces or parked cars in front of the known Soviet establishments. The best L.O.s, lookouts, prided themselves on recognizing KGB officers not only by face but by the way they walked when they emerged after dark. Mobile surveillance teams of specialized support personnel followed the hottest KGB players.

There were also special operations squads that broke into apartments, offices, and cars and planted bugs, ran wiretaps, and set up front businesses and checking accounts for undercover operations. In the early seventies, when the Soviets were building the Sovplex, the surveillance/technology unit, nicknamed "Susie" (easier to say than the original nickname "su-tech") had infiltrated construction crews and studded the compound with microphones. Agents from Susie and Division Three often worked together to plant listening and tracking devices in KGB cars.

Susie, whose official code name was MEGAHUT, was so secret that its members didn't work out of the field office, but rather in an unmarked office in Midtown, also frequented by officers from the National Security Agency, the Defense Department's supersecret communications intelligence agency that specialized in interception and code-breaking. Since the Church Committee revelations, it had been illegal for the NSA to conduct covert activities within the United States. So the FBI — Susie — planted the listening devices, and the NSA took the product, then translated, deciphered, and disseminated it within the intelligence community.

Hanssen was designated a case agent on Squad 32, covering KGB Line X officers, who specialized in classified scientific and technological information. Other FBI squads focused on Line PR (political intelligence), Line N (illegals), and Line KR (counterintelligence), which kept a watch out for FBI surveillances on other KGB officers and for potential defectors among the Soviet diplomatic corps.

In all, there were about 200 case agents in eight or nine Division Three investigative squads. The task of a case agent was to determine whether new Soviet arrivals were officers of the KGB or the smaller GRU (Soviet military intelligence), then decide which ones warranted a surveillance by Susie and the case agent himself. Each case agent was assigned a list of known and suspected Soviet intelligence officers to be studied in minute detail, by means of personal observation and investigation, through analysis of information from other human and technical sources, plus whatever the CIA, the NSA, and friendly overseas intelligence services had to offer. The case agent also evaluated every KGB officer on his list for possible recruitment purposes. Was he in financial trouble? (It would be a surprise if he were not.) An alcoholic? Was his marriage shaky? Extramarital affairs? Had he been passed over for advancement? Was he suspected of disloyalty, fairly or not?

In Spycraft 101, Hanssen would have been instructed to get to know a Soviet's daily habits to the minute. "You had to get a pattern on them, so you'd know when they broke their pattern," said Curran. A deviation from routine could mean the officer was heading out to service a dead drop or make a clandestine meet.

But while the pattern theory worked well enough in Washington and San Francisco, where the FBI outnumbered the KGB by a comfortable margin, there were simply too many Soviet spies in New York to watch. An around-the-clock surveillance of a professional intelligence officer, trained to "dry-clean" himself by changing subways, ducking in buildings, and driving evasively, required as many as five or six FBI personnel, and perhaps a light plane. "We were overwhelmed by the numbers," said Jean Gray.

Hanssen arrived in Division Three just as the most important counterespionage case in years was in full swing. The FBI had not discovered an important "illegal" since the KGB's Colonel Rudolf Abel was unmasked in 1957. But one bright Sunday morning in the spring of 1977, an FBI agent who lived in northern New Jersey had spotted a car with a Soviet diplomatic license plate. Division Three traced the plate to a car used by the chief of KGB Line N, illegals. There was only one reason the Soviet would be out that far in the suburbs — a clandestine meeting.

The Soviet was watched day and night for about a month. Finally, on Memorial Day Weekend, he drove out to Westchester County, the Soviet Mission's preferred location for drop sites. A car drove past him, and the teenage boy on the passenger's side tossed something out. The KGB officer retrieved the package and headed back to the city.

Agents following the other car learned that the driver was a man who called himself Rudolph Albert Herrmann, a Czech freelance photographer who lived in Westchester with his wife, Inga, and sons Peter and Michael. After investigating him for some months, FBI agents paid him a call, confronted him and threatened to arrest him and his son Peter, who had made the toss. Herrmann quickly agreed to cooperate with the FBI as an RIP, FBI-speak for "recruitment-in-place." His real name was Dalibar Valoushek, and he was a colonel in the KGB. Moscow Center had sent him to the United States to be a sleeper agent nine years earlier. Herrmann provided a wealth of inside information about KGB codes, secret writing, "accommodation addresses" — cutouts — and other covert communications methods to be used in the event of a break in contact. He also told them that Peter, who spoke German and Czech, was studying political science at Georgetown University and was being groomed by the KGB to join the U.S. Foreign Service as a long-term penetration agent.*

The FBI hoped Herrmann's capture would send a message to Moscow. But that didn't happen. Moscow Center became even more aggressive, upgrading its technical penetration gear. At the time, New York, Washington, and other major U.S. cities were replacing old-fashioned, hard-wired telecommunications equipment with microwave transmission systems. These were childishly easy for the Soviets to intercept. The San Francisco consulate's rooftop sprouted antennae aimed at U.S. military contractors along the coast and the infant computer industry in Silicon Valley. Antennae atop the Soviet Embassy on Sixteenth Street in Washington collected unscrambled signals emanating from the White House, the State Department, the Pentagon, and CIA headquarters in Langley, Virginia. An even better listening post was under construction. In August 1978, just as Hanssen was showing up to work at NYO, the U.S.S.R. broke ground on a new embassy on Washington's highest point, a 12.5-acre plot where the Mount Alto Veterans' Hospital once stood, on Wisconsin

---

*Acting on the FBI's instructions, Herrmann sent messages to the Soviet Mission seeking meetings but got no response. The FBI agents smelled a compromise; years later, Vasili Mitrokhim, the former KGB librarian who defected to Great Britain with a great quantity of documents and notes, claimed that Herrmann used secret signals in his messages that warned Moscow Center that the FBI was on to him. At any rate, in September of 1979, the FBI decided Herrmann had kept his end of the bargain and helped the family relocate under another name.

Avenue and Calvert Street, NW. Under a 1969 agreement, the Nixon State Department allowed the U.S.S.R. to acquire this choice location, also known as Mount Alto, in exchange for which the United States got the Kremlin's permission to build a new embassy in Moscow. Years later, intelligence community officials realized that the location facilitated the Soviets' abilities to intercept microwave telephone calls and faxes in U.S. government buildings and possibly even pick off high-level conversations by bouncing laser beams off the White House windows. The Soviets had also acquired high ground for its Riverdale residence by leveraging the Nixon administration's eagerness to make détente work. Now the Sovplex's rooftop dishes eavesdropped on communications throughout the Northeast.

Agents in New York and Washington smelled the change in the wind and intensified their quest for recruitments-in-place. RIPs were, said Jean Gray, "the Holy Grail of counterespionage work." Agents dreamed of becoming another John Mabey, who had recruited the legendary master spy TOPHAT. TOPHAT was the FBI code name for General Dmitri Fedorovich Polyakov of the GRU. The barrel-chested combat veteran with piercing eyes was the most highly placed, knowledgeable, and long-lived mole the U.S. intelligence community ever had — "the jewel in the crown," as James Woolsey, the director of Central Intelligence from 1993 to 1995, described him. Polyakov, a World War II artillery officer decorated for bravery, was a Russian patriot who despised the Soviet system. In 1961, while a forty-year-old lieutenant colonel on the Soviet Military Staff Committee of the Soviet Mission, he approached an American at a UN function and hinted he would like to volunteer his services. Mabey made contact with him and after some back and forth, they struck up an arrangement. A year after signing on with the FBI, Polyakov was posted back to GRU headquarters on the outskirts of Moscow. In 1965, he was promoted to full colonel and posted to Rangoon as the GRU resident. He lived there until 1969, meeting frequently with FBI and CIA personnel. In 1969 he went back to Moscow, where he headed up the GRU's China branch. In 1973, he was promoted to general and sent to New Delhi as the GRU resident. He returned to Moscow in 1976, then began a second tour in New Delhi in 1979.

Polyakov told the Americans with whom he met that he wanted to make sure the Soviets lost the Cold War. "I think his motivation went back to [World War II]," said one CIA officer who dealt with him. "He contrasted the horror, the carnage, the things he had fought for, against the duplicity and corruption he saw developing in Moscow. He articulated a sense that he had to help us out or the Soviets were going to win the Cold War, and he couldn't stand that. He felt we were very naive and we were going to fail." Also, Polyakov wanted to rise in the GRU so his two sons would be well educated and assured of professions. To that end, the CIA gave him some minor secrets and provided two Americans whom he could claim to have recruited. The two were actually double agents and reported back to the CIA.

Polyakov informed on several Westerners recruited by the GRU, including Frank Bossard, a guided-missile researcher in the British aviation ministry, Navy yeoman Nelson C. "Bulldog" Drummond, U.S. Army Sergeant Jack Dunlap, an NSA courier, and U.S. Air Force Sergeant Herbert W. Boeckenhaupt, who repaired code machines and sold secrets of the Strategic Air Command. James Angleton, the CIA's famously suspicious chief of counterintelligence, harbored suspicions that TOPHAT, also code-named DONALD, BOURBON, and ROAM, was a Soviet plant, but all doubts were set aside as Polyakov kept coming up with pieces of information too precious to Moscow to be throwaways. Whenever his reports arrived, recalled Sandy Grimes, a headquarters intelligence officer who handled information from Polyakov, "It was like Christmas." In the late sixties, at the height of the Vietnam War, Polyakov, then posted in Burma, gave the United States a bounty of data about the Vietnamese and Chinese military. Rotated back to Moscow, he purloined documents tracking China's bitter split with Moscow. Secretary of State Henry Kissinger used these data to forge the Nixon administration's 1972 opening to China.

After Polyakov was promoted to general in 1974, he got his hands on all manner of useful intelligence, including the coveted high-tech shopping list. Polyakov photographed more than a hundred issues of *Military Thought,* a monthly publication of the Soviet general staff. From these, CIA analysts learned that the consensus among Soviet military strategists was that they could never win a nuclear war with the United States. In 1979 or 1980 he told a CIA handler, "I think your military communications are penetrated." That was the first hint of the infamous Walker espionage case that would be uncovered in 1985. The secrets Polyakov stole would benefit generations of Americans; in the seventies, he passed along technical data on Soviet-made antitank missiles that would be deployed by Iraq during the 1991 Gulf War.

Polyakov's tradecraft awed the Americans who dealt with him. When posted in the Third World, he met FBI or CIA handlers, usually while fishing, but in Moscow he ruled out face-to-face meetings. Instead, he dipped into GRU supplies for self-destructing photographic film that could be developed only with a special chemical — normal photo chemistry would turn it blank. He deposited the film in fake rocks that he left in fields, to be picked up later by CIA officers. In the late seventies, CIA technicians invented a special, handheld device into which he typed information which was then encrypted and transmitted in a 2.6-second burst. He activated it while riding the tram past the U.S. Embassy in Moscow.

Polyakov never accepted more than $3,000 a year. His requests were simple: Black & Decker power tools for weekend carpentry projects, a pair of work overalls, fishing gear and shotguns, which he collected, and a few trinkets like lighters and pens — wonderful status symbols for Soviets — to be passed around to other GRU officers as favors. He did not drink or smoke to excess, was faithful to his wife, and spoke mostly about his children and grandchildren.

Hanssen's colleagues knew that they had about as much chance of finding another Dmitri Polyakov as they did of batting like Mickey Mantle. But they tried. An FBI agent couldn't just waltz up to a Russian and make him an offer. The Soviet might take offense and file a complaint with the State Department, which would subject the agent

to a blistering lecture for harassing foreign guests. Blackmail sometimes worked — not the crude honey trap favored by the KGB, but
subtler stuff. The FBI knew, for instance, that KGB officers sometimes
shoplifted small, pricey items like watches and cameras to sell back
home. The case agent could tip the store, materialize, and help the
Soviet out of a jam.

To make contact with other KGB officers who might be thinking of
crossing over — whether to solve financial woes or wreak revenge on
a sadistic boss or a thousand other reasons the FBI could not
know — the case agents would hit the streets. "You had to make sure
they knew you, if they wanted to talk to you, without overplaying
your hand," said Jerry Doyle, one of Hanssen's colleagues in Division
Three. "It was a courtship. You don't look too interested, but you're
there if they're interested. I don't know how many times I sat on
benches on the West Side, freezing my backside off, making sure if a
Soviet walked by, he knew I was his case officer."

Timing was everything. "If you followed them all the time, they
knew you weren't an officer," said Doyle. "You had to make it clear
you weren't just a follower. You showed up every so often. You'd dress
nicely, in a suit and trench coat. You'd read a newspaper and make eye
contact. They'd nod and acknowledge you. It was very intellectually
challenging work in which you really used your mind." If it
worked — if one fine day a KGB officer flung open his arms and sub-
mitted to the FBI man's tender mercies, said Doyle, "you were really a
star."

It was obvious early on that Bob Hanssen wasn't going to be a star,
not at the FBI. The raw IQ was there, but not the adroitness and
charm that marked a good recruiter. In school, Hanssen had usually
done the minimum — and that only if he was already interested.
Now his laziness revealed itself in his investigative efforts. "He didn't
get his hands dirty," said Ed Curran. "He was never out on the street
nights, weekends, holidays." He also didn't have much in the way of a
sense of humor. Many of his fellow agents found him just plain weird.
He never wore anything but black suits that looked as if they had

come from the Addams family's closet. He got right up in people's faces and whispered, as if everything was a conspiracy. Nor did it help that he had bad breath.

And, for a new guy, Hanssen sure seemed to think a lot of himself. "I always thought he was an arrogant, narrow-minded asshole," said one agent who worked alongside Hanssen. "You had to acknowledge that he was smarter than you. He was always looking down his nose at the Neanderthals working the cases." When he asked questions, it was in an odd way, reminiscent of the way he had dealt with Jack Clarke back in Chicago. "Cunning" was the word Clarke had used, and it still applied. It was as if every conversation had another purpose for Hanssen. It was all one way, too. He engaged in long, abstract, philo-sophical monologues concerning theology and communism. Bob saw no need to bond with his colleagues or to reveal much about himself. At times, it seemed as if he simply didn't see the point in get-ting to know lesser men. A colleague recalled that if a secretary was struggling with both arms full of file folders, Bob would never hop up to open the door. He never went out for a beer with the guys at the Sun Luck, the NYO watering hole; Hanssen made it very clear that his fervent Catholicism and membership in Opus Dei precluded such entertainment. There was a large contingent of Catholics in the office — Catholics by heritage, not conversion — and they didn't appreciate his exaggerated pieties. "People who are super-religious, and only God meets their standards, usually have no time for mere mortals," is how Doyle put it.

As had been evident since his ham radio days and from his investi-gation of the Soviet ship communications systems while in Gary, Hanssen was very good with anything electronic. Soon after NYO's April 1979 move from the Lincoln warehouse to a bigger, updated space in Foley Square, information technology experts commis-sioned by Director William Webster started setting up a database for FCI work called the Intelligence Investigative System (IIS). They con-templated that IIS would replace paper as a storehouse for names, addresses, associations, and other minutiae about the intelligence officers the FBI were targeting. Ultimately the experts hoped to give IIS an artificial-intelligence capability so that it could discern obscure associations among seemingly unconnected individuals.

Hanssen was not a computer programmer by training, but when he heard about the fledgling IIS project, he stepped forward to volunteer for its New York arm. His supervisors took him up on the bid and transferred him to a technical-support squad.

The culture of Division Three was fun-loving but also discreet. The guys ragged each other mercilessly, with plenty of laughs over what some mope had done out in Brooklyn, or how somebody had outmaneuvered a State Department desk officer, but nobody talked around the watercooler about real cases. "If you got a RIP, you'd tell everybody he blew you off, so interest would die down," said Doyle.

Hanssen didn't get it. He peppered supervisors and agents with questions. "I didn't like him, so I didn't answer him," said Doyle. "I used to say, 'What prompts your concern?' I just figured he was nosy, that he wanted to learn as much as he could about the work."

Having become the office computer nerd, Hanssen was ignored or just not noticed by numerous agents. Others dealt with him, but regarded Hanssen as mildly annoying. The guys started calling him "Dr. Death" and "The Mortician." "He reminded you of Ichabod Crane," observed Doyle. "If you saw him on a Halloween night, he'd scare you. He looked like someone who would lurk in the shadows." The epithet Dr. Death was also a reference to Hanssen's studies in dentistry — and to "Dr. Szell," the mad Nazi-war-criminal-dentist-torturer portrayed by Sir Laurence Olivier in the movie *Marathon Man*.

Hanssen didn't have much good to say about his partners, either. He confided in his childhood friend Jack Hoschouer that he had become disillusioned. "He figured that Russians were doing illegal stuff when [FBI] people were home in bed," recalled Hoschouer. "He said that guys had to get out on Sunday morning, but they didn't want to do it. He was disappointed that they were not as motivated as he was, that Russians got away with it because others were screwing off." It was true that not all the KGB officers were under tight surveil-lance. But it was also true that lots of agents were working nights and weekends. Hanssen just wasn't one of them.

As he had so many times before, Hanssen dealt with his role as an outsider simply by withdrawing further. Change was too hard, and after all, why should he be the one to adapt? "He was very literal-minded," said someone who knew him well. "He did not think well out of the box." So instead of reaching out, he toiled away at the com-puter, feeding in data from the thousands of index cards in a filing system known as MOSNAT (for "movement of Soviet nationals"). These cards represented a master index to all NYO Soviet counterin-telligence files. Essentially, they summarized what the FBI and the intelligence community had on every known and suspected Soviet intelligence officer, every Soviet diplomat who had been investigated as being "co-opted" by the KGB, and every American and third-country national who had ever come under suspicion as a possible Soviet agent. Each card bore a case number, name, address, job description, work and personal history, names of spouse and children, previous assignments, and a snapshot. Hanssen also entered into the system information from some of the case files themselves, plus every scrap of data any agent would entrust to him. Agents who worked in Divi-sion Three do not believe he was given the true names of RIPs and other FCI sources. These were, and still are, kept on paper, in safes. In reports, sensitive sources are given numbers and symbols. But he cer-tainly could have discerned something about a particular source's location and access by the kind of detail he or she was imparting.

All of this was way beyond what his father had had access to as a city cop — that was obvious. And it was Bob's kind of power, because all those guys who laughed at him, who called him names (yes, he knew what they said behind his back) didn't realize what he could do now. Intelligence and counterintelligence were all about getting and controlling information. Let those guys get drunk and stay out all night. Hanssen had hit pay dirt.

7

THE SPYING BEGINS

Over the months, Hanssen amassed a prize collection of secrets. He had all kinds of information about the capabilities of Division Three — how many agents there were, how many squads, how the squads were organized, the tag numbers of all the FBI cars. He had some biographical information about individual agents. He could make an educated guess at what the squads were doing at any given moment. Although the FBI supervisors tried to compartmentalize their work, whenever a big case was in the works, such as the operation that netted Rudolph Herrmann, the office would crackle with energy. Some agents would disappear for days, while others camped in the office nights and weekends. Supervisors ran upstairs two or three times a day to brief the ADIC or sit in on a conference call with HQ.

With his complete access to the New York office's Intelligence Investigative System database, Hanssen could see which Soviet intelligence officers had been identified by the FBI and which had not, what was known about them and, possibly, who was being monitored most closely. He understood that when a "lookout" agent famed for having memorized every known KGB face in Manhattan was moved from the "bucket" outside the Soviet Mission to Penn Station, it was because one of the squads had determined that certain KGB officers had stopped driving to their drops and were using the Long Island Rail Road. As a technophile, he knew more than most agents about Susie's capabilities and the listening devices planted inside the Sovplex, KGB apartments, and their cars.

Bob Hanssen had always seen himself as one of the chosen, even if others in their ignorance did not recognize that truth. Those who observed his growing involvement with Opus Dei could understand why being part of an elite bound for ultimate salvation might appeal to him. It was a world of right and wrong, black and white, while the life of counterintelligence was filled with shadows, a kaleidoscope made up entirely of shades of gray. And although Hanssen may have felt secure when it came to the metaphysical, real life in New York was a shakier proposition. A little house in Scarsdale was cheaper than a Manhattan apartment, but with four kids sucking up much of his salary, Hanssen faced an even more intractable dilemma than the other rookies, most of whom were younger than he and either unmarried or only recently wed. He and Bonnie were close to bankruptcy. Worst of all, once again Hanssen found his intellect unappreciated. He had not managed to shake the bad memories of his dad, and now his fellow FBI agents, when not tracking spies, seemed to get a kick out of coming up with cruel nicknames for him. Sure, he never went out boozing with them. But did they ever invite him? Did they ever even notice him? Did anyone?

Hanssen was a smart guy, and it had to eat away at him when others made fun of him, laughed behind his back. He didn't know what they spoke about when out at the bars and strip clubs, but maybe it was about him — maybe that was all they ever talked about. Hanssen had spent his life pressing his face up against the glass. Too arrogant to work his way up, too self-righteous to socialize with those he considered his inferiors (that is, everyone else), more than anything he wanted power — *real* power, which to Bob meant the power to humiliate others. The damsel-in-distress fantasies that Jack Hoschouer had noticed in high school were always as much about reminding the girl that she was totally dependent on Bob as about helping, maybe more so. Hanssen had enjoyed the C-5 back in Chicago because it gave him the power to end careers, not because it gave him the power to clean up the force. And if his coworkers at the FBI didn't realize who was *really* boss, well, Bob knew one way to splash some cold water in their faces. And he was so smart, so clever, that even after drying off, they wouldn't know who had doused them.

One day in 1979, Hanssen walked into the Amtorg office in Manhattan. He knew that Amtorg, which ostensibly brokered trade deals on behalf of the Soviet government with American companies, was actually an important GRU front for collecting scientific and technological information for the Soviet military. Hanssen gave a phony name and left a message to be delivered to a GRU official. It was an offer to sell classified government information.

His life as a spy had begun.

The first thing Hanssen betrayed was the identity of Dmitri Polyakov, aka TOPHAT. Hanssen knew very well what happened to traitors in the Soviet Union — they were lucky to be shot outright — but he knew that for pros, spying is a blood sport. The bigger the prize, the more respect he'd get. Getting rid of Polyakov might actually protect Hanssen, too. The GRU was supposed to be highly compartmentalized, but Polyakov was a general, with access to just about anything he wanted. Although Hanssen went to great lengths to conceal his identity, the GRU could have figured it out it by staking out a dead drop, tailing him home after he picked up the money, and tracing the owner of the house through property records and a crisscross directory. If Polyakov had the source's name, he would let the Americans know, and that would be the end of Bob Hanssen. So, from Hanssen's point of view, it was a matter of Darwinian survival. He would get TOPHAT before TOPHAT got him.

Hanssen also knew his chances of getting away with it were excellent. Although the FBI initially recruited Polyakov, the CIA had taken over handling him in the mid-1960s. If he were executed and a mole hunt ensued, suspicion would naturally fall on the CIA officers who had dealt with his case, and as there were so many of them, the investigation would grind to an inconclusive halt, without ever getting around to the FBI.

The GRU did recall Polyakov from New Delhi in March of 1980, but he was not arrested. Possibly top GRU officials did not believe Hanssen's information because his bona fides had not yet been established. Possibly they suspected Hanssen of being an agent provocateur, dispatched by U.S. intelligence to sow discord among them by passing disinformation about traitors within the GRU's top ranks.*

It was Hanssen's wife who first noticed something was wrong. One day in early 1980, Bonnie went to the basement of their home and found Bob doing something that struck her as strange. People close to the Hanssens have heard different versions of the incident — one, that he was working on some papers that he tried to conceal; another, that Bonnie suspected him of having an affair. Everyone agrees that Hanssen told his wife only half the truth: that he had forged a deal with the Soviets. He insisted that he hadn't given them anything important, just some useless information, trash for cash. And a lot of cash at that, as much as $20,000. Bonnie, horrified, begged her husband to consult a priest with her. Hanssen also consulted a lawyer, but the advice he received is not known.

Hanssen, responding to his wife's entreaties, confessed his espionage to a Catholic priest. First identified by the *New York Times*, the Reverend Robert P. Bucciarelli acknowledged only that he knew Hanssen in 1980. Bonnie told authorities that Bucciarelli urged Hanssen to turn himself in, then changed his mind and instructed Hanssen to

---

*Not surprisingly, the GRU appears not to have alerted the KGB. The two organizations tolerated each other, but that was all. GRU officers were proud that all their spying took place overseas, for the sole purpose of benefiting the U.S.S.R.'s national defense. They disdained the KGB for its internal secret police branch, which intruded into the lives of Soviet citizens and used blackmail and intimidation to repress antigovernment sentiment.

give his ill-gotten gains to charity; Bonnie claimed Bob sent the money to Mother Teresa. This story made sense to a Hanssen friend familiar with his anticommunist philosophy: "Bob probably thought he put one over on Brezhnev, that he took their money and gave it to charity," he said.

But it didn't make sense to other priests. Father Bucciarelli, chaplain of the Opus Dei center in Cambridge, Massachusetts, is a major star in the Opus Dei constellation. From 1966 to 1976 he was the head of Opus Dei in the United States. While priests never discuss what advice they give in confession, Father Bucciarelli's friends insisted that such an experienced priest would most likely counsel Hanssen to turn himself in, not to give the money to charity. "That's the right answer," said Monsignor William Smith, professor of moral theology at St. Joseph's Seminary in New York, the seminary of the Archdiocese of New York. "Even if you are not in confession, you have to tell him he has to make restitution and the price. He has to make reparations. If he has caused unjust harm or other people's loss of life, I would not absolve him until he promises to do that."

Most people would understand why a woman with four children, all totally dependent on her husband's income, might remain silent, if only for the sake of financial survival. But did the priest have a responsibility to report a felony and, perhaps, prevent future acts with more serious consequences?

Under the Code of Canon Law, which prescribes the laws of the Roman Catholic Church, a priest may not disclose to anyone any information learned in confession. "The sacramental seal is inviolable," the law says. "It is absolutely wrong for a confessor to betray the penitent for any reason, whether by word or any other fashion." Anyone who remembers Alfred Hitchcock's 1953 movie *I Confess*, starring Montgomery Cliff and Anne Bancroft, knows that a priest is bound to protect the confidence of the sinner, however dire the sin, whether it be treason, adultery, or murder. In Hitchcock's flick, Otto the gardener confesses to Father Logan that he has committed murder, but even after the police suspect Father Logan to be the killer and put him on trial, the priest does not reveal the truth.

In the half-century since that movie was filmed, nothing has changed. "The priest's job is not to turn someone in," said Opus Dei's Father C. John McCloskey III. "He is not a cop or a judge but an instrument of God's mercy."

Father Robert Drinan, an ordained priest and former five-term congressman from Massachusetts as well as a professor at Georgetown School of Law, illustrated the point with a hypothetical case used in seminary classes: a penitent confesses his sin, receives absolution, and then tells the priest that he has poisoned the wine that the priest will serve at mass the next morning. "The priest has to go to mass and use this wine," Father Drinan said. "It shows the absoluteness of this bond." Canon law, he explained, is very explicit: "The confessor is wholly forbidden to use knowledge acquired in confession to the detriment of the penitent, even when all danger of exposure is excluded."

Not everyone takes Drinan's hard line. The Reverend John Beal, chairman of the Department of Canon Law at Washington's Catholic University, offered a compromise solution: "He could throw out the wine as long as he didn't do it in a way that called attention to the penitent," Father Beal said. "If the priest quietly walked into the church, smelled the wine, and said it had gone bad and thrown it down the drain, he would not attract any attention to the penitent. But he could not go to the authorities and have him arrested for attempted murder."

The situation is somewhat grayer if a priest is in a counseling session when he learns that a person has committed a criminal offense, as may have been the case with the Hanssens. In these situations, a priest tries every way he can think of, short of alerting government authorities, to get the person to correct the situation. "Your first line is: 'Can you release me from this?'" said Father Smith. "If that fails, your next line is: 'Can you tell someone else? Can you bring it out of this forum so we can solve this problem honorably, because you really are a stinker.' If he won't, then basically I'm stuck."

Smith added that if he had heard the information in a counseling session with a husband and wife, he would do as much as he could within canon law to make the information public. "I would force him to release me or I would make his situation so difficult," he said. "You lean on the wife, you lean on her hard. You tell her that the husband has to straighten himself out. You tell her to inform her husband that there is only one correct moral solution and that he must report himself to the government and undo any damage he has done." If anyone leaned on Bonnie Hanssen, no one knows. Asked to explain what happened, Father Bucciarelli replied, by e-mail: "A priest must decline to give information on confidential conversations, so it would be inappropriate for me to say whether I ever met with Mr. Hanssen in the context of spiritual consultation. I have no other comment than to suggest you might consult a professor of moral theology." He would not confirm or deny that what had been written about him in the *New York Times* was correct.

Hanssen turned to spying so soon after joining Opus Dei that some within the community believe he was faking his religious commitment from day one, that he embraced Opus Dei in a cynical, calculated bid to create a perfect cover, or "legend," as they say in spycraft. But people are complicated, especially unhappy people. Virtually all who met Bob Hanssen in the twenty-five years between his conversion to Catholicism and his arrest considered him to be a man of faith, a devout Catholic and staunch anticommunist to the innermost reaches of his soul. His priests did, too. But Hanssen had a way of being seen but not seen. Just as his fellow agents would head out for a drink, oblivious to any hints that the man they called "Dr. Death" was dangerously abnormal, Hanssen's priests seem to have noticed him in the back row but never devoted deeper attention. Over the years, no priests really monitored Hanssen's spiritual direction. Three priests volunteered that while Hanssen had attended mass at their churches on a regular basis, he never went to them for confession. None knew Hanssen personally. Two had no more than a nodding acquaintance. Father John O'Neill, pastor of Our Lady of Good Counsel, said Hanssen had been one of the thirty-five to forty regulars at his 6:30 A.M. mass for at least the past decade. O'Neill said that if a parishioner like Hanssen had come to him with his confession, he would have done all he could to help. But perhaps the FBI wasn't the only institution Bob Hanssen was fooling.

Whatever the truth, one thing seems clear: over twenty years before he was arrested for causing more damage against the United States of America than any FBI agent in history, Hanssen had admitted his involvement with the Soviets. Nobody did anything about it. That lack of response may have encouraged Hanssen in all sorts of ways. It may also have encouraged the Soviets. And given Hanssen's religious faith, with its insistence on true absolution, a stark question remains: If Hanssen did confess his espionage and no one intervened, who betrayed whom?

8

ASSIGNMENT: WASHINGTON

In January 1981, Hanssen received a transfer to Washington. His savior was FBI headquarters, which drafted him for the Intelligence Division's budget unit. The job required him to prepare budget proposals for submission to the intelligence community staff, the Office of Management and Budget, and the congressional appropriations committees. Any field agent worth his wingtips would have considered himself doomed to Devil's Island, but the budget unit's desultory pace suited Hanssen just fine.

Besides, if you're a know-it-all, there's no better place to know it all than the budget shop. No matter how compartmentalized individual counterintelligence programs are, the wraps have to come off when the time comes to pay for them. Hanssen was able to zoom in tightly on an individual program, then zoom way out for a panoramic view of the U.S. government's entire counterintelligence effort, not just the FBI's programs but those of the CIA, the NSA, the Defense Intelligence Agency, and other parts of the intelligence community. "All those programs were put together in one document for our budget presentation," said Pat Watson, who was a unit chief in the Soviet section at the time. "The document gave the view at a fifty-thousand-foot level, but he had to have access to a lot of information to assemble it."

The move was good for the Hanssen family finances, as well as its morale. Bob and Bonnie bought an attractive red and white brick house on Whitecedar Court in Vienna, Virginia, large enough to accommodate a growing family. They enrolled Jane (age nine) and Susan (seven) in the elementary school affiliated with Our Lady of Good Counsel, their local parish. Jack was four, and Mark, who had been born in April 1980, just fifteen months.

Hanssen couldn't have picked a better moment to arrive in Washington. He had a ringside seat on one of the most pivotal moments in the history of the secret spy war against the Soviet Union. President Reagan had been sworn in on January 20, 1981, and he was an unabashed cold warrior who turned up the heat on what he called "the evil empire." One of Reagan's first acts was to have his staff review U.S. counterintelligence efforts and, on December 4, 1981, he signed Executive Order 12333, directing a vast expansion of them. Attorney General William French Smith used the occasion to denounce what he termed the "dramatic" increase in spying by the Soviets and other hostile nations, who, he said, had expanded the number of intelligence officers in the United States by 400 percent in the last dozen years. The executive order was welcome news at the FBI, which stood to gain more people, more money, and a lot more respect from the White House and State Department — from the bureau's point of view, not a moment too soon. The FBI estimated that the KGB's First Chief Directorate, responsible for spying on the United States and Canada, had 6,000 agents overseas, plus increasingly powerful electronic interception equipment. There was a new breed of KGB man, no longer the stereotypical Vodka-swilling lout in a baggy suit and cardboard shoes, but rather a youngish, hip-looking university graduate, who spoke several languages and had a strong background in the sciences. KGB officers could often be found knowledgeably chatting up scientists and engineers at seminars, trade shows, and academic conclaves.

The need for the spies was clear: the U.S.S.R.'s indigenous military research and development programs were simply unable to match the technology revolution exploding in the West, and they were falling further behind by the hour. Around the time the Reagan administration took office, Polyakov gave his CIA contacts a shopping list of technologies sought by the Soviet military. According to Richard Perle, an assistant secretary of defense in the Reagan administration, the document, several inches thick, astonished even his fellow hawks with its meticulous detail and staggering scope; it showed that some five thousand separate Soviet programs depended on stolen Western technology. Perle used this information to convince Reagan to give the Pentagon the power to block overseas sales of technologies that could modernize the U.S.S.R.'s military capabilities. Reagan's arch-conservative U.S. customs commissioner, William Von Raab, launched Operation Exodus to counter rampant smuggling of high-tech exports to the Soviet Bloc through false-flag (innocent-looking) fronts in Europe and Asia.

In 1982, anti-Soviet hardliners, then in ascendance on Capitol Hill, approved FBI director Webster's five-year plan to double the number of foreign counterintelligence (FCI) agents. Also that year, Congress created a new Office of Foreign Missions (OFM) at the State Department, with veto power over any foreign government's purchase or lease of property within the United States. This provision, pushed by the FBI, NSA, and CIA, was aimed at preventing the Soviets and other hostile states from acquiring more properties that enhanced their abilities to spy electronically; unfortunately, as the U.S. intelligence community well knew, the Soviets had already nailed down two of the best technical listening posts on the East Coast, the new embassy site at Mount Alto, near the National Cathedral in Washington, and their residence in Riverdale, New York. The law required all nations that restricted the movements of U.S. diplomats — meaning, the Soviet Union, most of the Eastern Bloc, China, and Cuba — to make all their travel reservations through OFM. Areas around U.S. weapons laboratories and military bases, testing ranges, and certain other sensitive facilities were out of bounds. Congress did not have much faith in the State Department's ability to control the KGB, especially with the first inklings surfacing that the department had failed to prevent the U.S. Embassy under construction in Moscow from being honeycombed with KGB listening devices and transmitters. The White House tapped James Nolan, an expert in Soviet espionage, who was then head of operations for the FBI Intelligence Division, to be the first OFM director, with marching orders to clear all sensitive decisions involving the Soviets, China, and other hostile nations with the FBI and the rest of the intelligence community.

From his vantage point in the budget office, Hanssen was ideally positioned to know just about every aspect of the administration's counterespionage buildup. Moreover, his affinity for all things electronic earned him a spot on the FBI's FCI Technical Committee, which coordinated the insertion and monitoring of national security bugs, taps, and other tracking devices inside Soviet diplomatic offices, cars, and homes across the United States. He would likely have learned of the FBI's efforts to install tiny interception and transmitting devices inside the Soviet Embassy's Xerox machines, and also about cutting-edge eavesdropping devices beamed at the embassy from the outside. His access was not limited to anti-Soviet surveillance measures but included information about technical penetrations of the embassies of the Warsaw Pact nations, the People's Republic of China, Cuba, Vietnam, North Korea, Syria, Libya, Iraq, and Iran. And while Hanssen was sitting on the technical committee, the most ambitious and costly technical penetration the United States had ever attempted was taking place: the building of a tunnel beneath the Soviet compound on Mount Alto. The NSA funded the project, but because U.S. law forbade the agency to "go operational" on U.S. soil, the FBI's Washington field office was put in charge of the big dig. The bureau bought a house near the embassy and hired a contractor to burrow from the basement. Amazingly, there were no leaks, although countless tons of dirt were removed from a neighborhood that was home to some of the biggest names in government, law, and journalism. The NSA was less successful in its search for electronics that could penetrate the Soviet complex above the tunnel. Even so, U.S. counterespionage officials held out hope that the subterranean structure would someday prove its worth, which was mounting into the hundreds of millions of dollars.

In August 1983, Hanssen moved into the Intelligence Division's Soviet analytical unit. It was a lateral move, another desk jockey slot well removed from the sexy cloak-and-dagger world of FCI street agents, but he didn't care to view it that way. Years later, in an e-mail to his Taft High School alumni bulletin, he boasted of his success in the FBI: "My dad, before he retired, was District Commander of the Chicago Police Department's 16th District covering Norwood Park until heading up intelligence operations for the CPD downtown. He died several years ago. So, in the end, I became indeed a Special Agent with the FBI. I got promoted to Washington and headed Soviet Analysis at FBI headquarters in the Intelligence Division (like father, like son) during the Cold War." Once again, Hanssen exaggerated the facts to enhance his own accomplishments and the family status. Howard Hanssen was never a police district commander; he retired as a lieutenant. And Bob Hanssen never headed the FBI's Soviet analytical unit. He was one of two supervisory special agents overseeing sixteen analysts, reporting to the real head of Soviet Analysis, Tom Burns, unit chief from 1983 to 1987. Others got the job after that, but never Hanssen. Bob was at best a sergeant or lieutenant to the unit chief's captain. Few if any Taft alumni would have known the difference, or cared, but Bob Hanssen did.

Yet even cranky old Howard Hanssen would have been impressed by the treasures that crossed his son's desk. Every FBI case involving Soviet espionage anywhere in the world was sent to Soviet Analysis. The analysts had access to classified estimates from the CIA, NSA, DIA (Defense Intelligence Agency), and the State Department's Bureau of Intelligence and Research. Compartmentalization broke down in the analytical unit, as it had in the budget unit. The job of the unit was to examine all source information in great breadth and depth, looking for patterns, trends, and common players. Field agents working a case against Ivan Ivanovich in New York didn't need to know that agents in Los Angeles were working on Boris Borisovich, but headquarters needed to know about both of them — especially if Ivan and Boris turned out to be the same man, and he was hitting up defense contractors on both coasts for the same missile guidance system component, and had done the same thing in Paris three years earlier.

"All the field investigations come together at FBI headquarters," said Tom Burns. "It's the broadest possible view of what is going on at any given time across the country. It's just a snapshot, but when you put the snapshots together over time, you come up with an almost real-time understanding of what efforts are being made by foreign intelligence services." Ultimately, the FBI's analytic product was meant to give the director and his assistant director for intelligence the information they needed to define the nature of the threat and to design countermeasures.

Burns had seen the worst in people when he worked white hate crimes in the South during the civil rights era. (Among other things, he helped bring three Alabama Ku Klux Klansmen to justice for the 1965 murder of civil rights activist Viola Liuzzo.) Yet he still looked for the good, and he found much to like in Bob Hanssen. "He was very quiet, introspective, introverted," Burns recalled. "Whatever he did, he put a lot of mental energy into it. I thought he enjoyed the business of counterintelligence and always attempted to get involved in anything of significance."

Yet Burns, a fellow Catholic, who also worshiped at St. Catherine's, found Hanssen's convert zeal somewhat unsettling. And there was something else, something he couldn't put his finger on. "It was difficult to identify where his center of gravity was," Burns said. Hanssen had a way of creating an aura of drama around himself, as if his involvement in the spy world was real and not the vicarious life of a four-times-removed member of the pocket-protector set. "Bob spoke in a low voice, almost a whisper, in a manner that seemed to be intended to generate attention," said Burns. When speculating about why the KGB had done something, said Burns, most people in the unit tended to tiptoe around the edge of what was known, which usually wasn't much, and hedge with a lot of on-the-other-hands. One person might say he thought this, and another might venture that she suspected that, and so on around the room until it came to Hanssen. He would cock his head, pause for effect, and say, in a voice so soft that everyone had to lean forward, "Let me tell you what I understand as to what the actual facts are." "You got the sense," said Burns with a bemused smile, "that he believed he was the expert in many areas, that he knew all the details."

Given his fascination with cutting-edge technologies, Hanssen's imagination must have been fired by field-office reporting of aggressive KGB and GRU efforts to penetrate the Pentagon, U.S. defense industries, and civilian innovators of so-called dual-use technologies with both civilian and military applications, such as high-speed computers and lasers. In early 1982, FBI agents in the Washington field office got into a high-speed car chase with Major General Vasily I. Chitov, the Soviet Embassy's highest-ranking military attaché and a suspected GRU officer. When run to ground, he was found to have classified documents related to weapons technology in his car, after which he was declared persona non grata and expelled. In April 1983, Chitov's successor, Lieutenant Colonel Yevgeny N. Barmyantsev, also a GRU officer, was caught by the FBI attempting to retrieve stolen American military secrets from the base of a tree in rural Maryland. He was kicked out of the country. At about the same time, Oleg V. Konstantinov, a thirty-three-year-old KGB officer assigned to the Soviet Mission in New York, was arrested by FBI agents on his way to collect classified aerospace plans from an American who worked for a Long Island defense contractor and who was reporting the contacts to the FBI. In September 1983, Canadian authorities working with the FBI booted out two suspected KGB officers for attempted theft of military high-tech; one worked for a Soviet trade mission, the other, for the international secretariat of the United Nations' International Civil Aviation Organization.

In May 1985, the FBI rolled up the most damaging espionage ring since World War II, arresting John Walker, a retired Navy warrant officer and communications specialist, his brother, Arthur, a retired Navy lieutenant commander, John's son Michael, a Navy seaman/clerk-typist, and their friend Jerry Whitworth, a retired senior chief petty officer. Ringleader John Walker had been selling the Soviets highly classified naval communications and cryptographic information since 1967. This had allowed the Kremlin to intercept the communications of much of the U.S. fleet during the Vietnam War and to understand how the United States tracked Soviet submarines. The Navy's damage assessment concluded that the Walker ring's activities "permitted the Soviets to gauge the true capabilities and vulnerabilities of the US Navy. . . . Soviet access to [naval] operations and capabilities provided them with the motivation to dramatically improve the Soviet military posture and identified the specific steps which could achieve the largest gains relative to the US. It allowed them the focused insights required to reduce their own vulnerabilities, while simultaneously increasing the vulnerability of the US. . . . The Soviets were able to monitor the US Navy transition to the use of satellite systems as its principal communications network."

Not two weeks later came an unprecedented intelligence coup that brought with it a series of humiliations. On August 1, Vitaliy Yurchenko, the highest-ranking KGB defector ever, presented himself at the U.S. Embassy in Rome. Yurchenko, a top officer in the KGB's First Chief Directorate, had spent five years in the embassy in Washington and was now a key figure in the KGB branch responsible for external intelligence, counterintelligence, and "active measures," meaning covert actions such as disinformation campaigns. The CIA flew Yurchenko to Washington to debrief him. He immediately disclosed that the KGB had recruited an American, code-named "ROBERT," who had fingered several moles inside the Soviet bureaucracy who were reporting to the CIA. The code name was just a coincidence; Yurchenko was not talking about Hanssen. Instead, an urgent investigation established that "Robert" was Edward Lee Howard, a CIA case officer fired in 1983 for drug and alcohol abuse, theft, and mental instability. Incredibly, just before he was dismissed, Howard had been trained to go to Moscow as a case officer, which meant he knew all about the double agents being worked by Moscow station at the time. Howard's betrayal was one of the most damaging episodes in CIA annals, but Yurchenko's word alone was not sufficient to justify his arrest. The FBI set about making a case against him, meanwhile setting up a tight surveillance on his residence in Santa Fe. But on September 21, he gave the surveillance team the slip and fled to Vienna. Later, he showed up in Moscow, boldly giving interviews about his transgressions.

A second KGB asset identified by Yurchenko was Ronald W. Pelton, a forty-four-year-old former NSA communications specialist. He was arrested on November 25 and, a few months later, pleaded guilty to divulging highly classified communications intelligence sources and methods, including the fact that U.S. submarines had tapped into Soviet communications cables under the Pacific Ocean. Congressional critics were especially infuriated to learn that Pelton had met with Soviet spies at least twice in Vienna, a notorious KGB stomping ground, and on one occasion had stayed at the home of the Soviet Ambassador to Austria — all under the nose of the CIA's Vienna station.

Yurchenko, meanwhile, pulled his own disappearing act. On November 2, while dining with his CIA handler at a popular Georgetown bistro named Au Pied de Cochon, he excused himself and faded into the night. Two days later, he appeared at a news conference in Washington, D.C., staged by the Soviet Embassy and announced that the CIA had kidnaped and drugged him. CIA officials blamed a failed love affair for Yurchenko's change of heart, but ever after, counterintelligence officials would debate whether that was all there was to it, or whether Yurchenko's defection had always been a KGB distraction to cover up real mole tracks. Perhaps by giving up some moles that had run their course, the KGB could divert attention from those who were still useful.

By coincidence, two more espionage cases came to light in the month of November 1985. Jonathan J. Pollard, a Naval Investigative Service analyst, and his wife, Anne, were accused of spying for Israel, and

Larry Wu-Tai Chin, a retired CIA employee, was charged with spying for China for more than three decades. Pollard, who had given the Israelis a vast array of documents that identified extremely sensitive NSA sources in the Arab world, was sentenced to life in prison. Chin committed suicide.

After months of sensational headlines, it didn't seem possible that either side had anything left to hide. But in fact, each spy service was still sitting on some even bigger secrets. On the American side, there was KGB Lieutenant Colonel Boris Yuzhin, who for several years had been the FBI's eyes and ears inside the KGB office at the Soviet Consulate in San Francisco. Through Yuzhin, the FBI was able to monitor Soviet efforts to infiltrate the California academic community and the fast-growing information technologies industry around San Jose and Los Angeles.

Yuzhin, a semiconductor engineer by training, signed on with the KGB in 1967. He was one of fifty Russian scientists who came to the United States in 1975 as part of an academic exchange. Yuzhin went to the University of California at Berkeley, where KGB strategists figured he would have no problem recruiting assets in the leftist student body. Instead, Yuzhin himself was enchanted by the new freedom he enjoyed and immersed himself in books by Russian dissidents that were banned in the Soviet Union.

"I came to realize that all my knowledge, all my ideology was based on a lie," he would testify years later. Rather than defect, Yuzhin resolved to try to change the Soviet system from inside. He contacted a man he suspected, correctly, of being his FBI case agent and agreed to return to Moscow as a RIP.

In 1978, Yuzhin was back in the United States, this time as a San Francisco–based correspondent for the Russian news service Tass. He was rotated back to Moscow in 1982. According to author David Wise, Yuzhin led the FBI and Norwegian authorities to Arne Treholt, a Norwegian diplomat who had been turned by the KGB while posted at the UN. Treholt was arrested in Oslo in 1984.

When Hanssen arrived in Soviet Analysis, the FBI field office in Washington had just achieved its first-ever penetrations of the KGB residency in the Soviet Embassy. Operation COURTSHIP, as it was called, began in 1981, when D.C.-based FBI agent Bill Mann, posing as a defense contractor, met KGB Lieutenant Colonel Valery Martynov at a technical seminar. Martynov, an engineer in his mid-thirties, worked for KGB Line X (science and technology); his assignment was to troll for Americans with access to weapons-systems know-how, but in truth, he was fed up with the KGB bureaucracy in particular and the Soviet system in general. Mann gave Martynov his business card, made a lunch date, and began the recruitment dance. "It turned out that Valery was indeed looking for someone on the American side in whom to confide," Mann said. "But knowing how exposed double agents in the past had been tortured and executed by the Soviets for whom he worked, he was very hesitant in showing his hand until he could be assured that his safety, and that of his family, would never be jeopardized. It was my job to gain that trust." Over months of dinners and meetings, Mann convinced Martynov that his family's safety would be assured and "his identity would never be made known to more than a small circle of American associates." Martynov became a RIP in April 1982.

In January 1983, COURTSHIP scored a second coup. Agents in the Washington field office found out that KGB Major Sergei Motorin, assigned to Line PR (political) at the embassy, had a weakness for prostitutes and was surreptitiously acquiring stereo equipment to sell on the black market in Moscow. Using these minor transgressions as leverage, agent Mike Morton recruited Motorin, who gave the FBI the names of all the KGB officers in the embassy. Motorin was rotated back to Moscow in 1984 and continued to report to the CIA periodically.

Martynov remained in the United States, telling the FBI about various KGB efforts to steal strategic scientific and engineering data. His information enabled the FBI to alert U.S. defense contractors who produced or used specific technologies and provided intelligence analysts with an ongoing understanding of the shortcomings in Soviet weapons systems.

A large, gentle, jovial man, Martynov was devoted to his wife, Natalya, and their son and daughter. "He was very concerned about a medical problem his son had which required hospitalization," said Mann. "He had no confidence in Soviet medicine and was afraid his son might be butchered on a Moscow operating table." The Soviet Embassy refused to pay for treatment in Washington. Mann tried to think of ways to get the boy to an American hospital, but he could not, not without tipping the Soviet Embassy's Line KR (counterintelligence) team that Valery had contacts he could not explain.

Like Polyakov, Martynov had the charm, polish, and intelligence to rise to high ranks in his service. "Valery was a gentleman in every respect," said Mann. "He had a timid yet pleasing manner in speaking with people, was soft-spoken, smiled often, and always was considerate of others. He never forgot to express his gratitude to those who were kind to him. On one occasion, he brought flowers to a lady, who had invited the two of us into her home for dinner. He laughed often, and enjoyed hearing funny stories about other people and telling funny stories about people he knew. I don't ever recall Valery being in a bad mood during my year-and-a-half association with him."

No one would ever say any of this about Bob Hanssen. Although Tom Burns praised him as "diligent and thoughtful," his scowling expression discouraged camaraderie. Craving accolades, Hanssen could not have been happy that the analysis shop, where he excelled, was low in the FBI's pecking order. The awards and promotions went to guys and gals in the field who made big cases. So in mid-1985, Hanssen sought a promotion to field supervisor. The field supervisors were the heart of the bureau. They ran the squads that made the cases that made the FBI's legendary reputation. Also, if he wanted to make it to the top-paying jobs in the senior executive service, Bob had to start climbing the ladder. He was having trouble paying the bills. Bonnie had given birth to their fifth child, Greg, and now she was pregnant with their sixth. She believed in him. He wanted to make her proud, give her things, dress her up. He had seen how his friends envied him when she was on his arm. In April, Jack Hoschouer and some military friends, on a training mission near Washington, met Bob and Bonnie for dinner at the Old Ebbitt Grill, a popular Washington eatery, located just a block from the White

House. Hoschouer recalled that Bonnie was so radiant that when she walked into the restaurant, his classmates' jaws dropped. "I will never forget how wonderful Bonnie looked," Hoschouer said. "She was wearing a red dress and her hair, which Bob liked her to keep long, was up in a whirl on her head. She was a knockout."

There were only three field offices with important counterintelligence units: Washington, New York, and San Francisco. Hanssen found out about an opening for a position as supervisor of a technical squad in the New York Intelligence Division and he grabbed it. He knew from experience that the commute would be awful, the cost of living worse. But Hanssen also knew that nobody in the FBI became a senior manager until he or she had been a successful field supervisor. His family would just have to sacrifice for a while. Life was full of compromises.

He reported to New York on September 23, 1985.

9

THE LETTER

Moving once again, Bob and Bonnie Hanssen found a cramped three-bedroom, one-and-a-half-bath ranch house on Mead Street in Yorktown Heights, in Westchester County. Bob's commute took ninety minutes on a good day. With five children and one on the way in just a few weeks, his paycheck was going to be stretched thin.

Hanssen had promised Bonnie he would have nothing more to do with the Soviets. But he had also promised to take care of her for better or for worse. At the moment, things were worse. His bills were bigger than they ever had been. Then again, so were the secrets in his head. If the FBI cared all that much about these secrets, why were they so easy to get? If he didn't cash them all out, sooner or later somebody else would — if they hadn't already. The entire U.S. government was a sieve, and the FBI, for all its vaunted pride in its records system, had not done what should have been done to compartmentalize the intelligence databases. What he couldn't lay his hands on for his analytical work, he could get unofficially, by browsing the computer or just asking questions. Had everybody in the

Hoover Building been on Mars? It was, as *Time* dubbed it, "The Year of the Spy." Didn't that mean something?

As with any other crime, any sort of treachery, the second time you do it it's a lot easier. Once Hanssen resolved to rekindle his relationship with the KGB, he acted with bold, sure strokes.

Ten days before Lisa, his sixth child, was born, Hanssen mailed the letter. It was Tuesday, October 1, 1985. He had been back at headquarters on a few "administrative matters," as the FBI put it. As he drove north, toward New York, he left I-95 briefly to find a postal box. The note arrived three days later at the Alexandria, Virginia, home of Viktor M. Degtyar, the press secretary of the Soviet Embassy in Washington. Degtyar was a KGB Line KR (counterintelligence) political officer. Inside the letter was a second letter marked "Do Not Open. Take this envelope unopened to Viktor I. Cherkashin." Viktor Cherkashin was the Line KR chief at the Soviet Embassy and one of the most aggressive, able KGB officers around. The letter-within-a-letter gambit was a cunning move that had been used by other spies, and Hanssen had probably heard about it during office bull sessions.

Hanssen's letter, typed and unsigned, read as follows:

DEAR MR. CHERKASHIN:

SOON, I WILL SEND A BOX OF DOCUMENTS TO MR. DEGTYAR. THEY ARE FROM CERTAIN OF THE MOST SENSITIVE AND HIGHLY COMPART-MENTED PROJECTS OF THE U.S. INTELLIGENCE COMMUNITY. ALL ARE ORIGINALS TO AID IN VERIFYING THEIR AUTHENTICITY. PLEASE REC-OGNIZE FOR OUR LONG-TERM INTERESTS THAT THERE ARE A LIMITED NUMBER OF PERSONS WITH THIS ARRAY OF CLEARANCES. AS A COL-LECTION THEY POINT TO ME. I TRUST THAT AN OFFICER OF YOUR EXPERIENCE WILL HANDLE THEM APPROPRIATELY. I BELIEVE THEY ARE SUFFICIENT TO JUSTIFY A $100,000 PAYMENT TO ME. I MUST WARN OF CERTAIN RISKS TO MY SECURITY OF WHICH YOU MAY NOT BE AWARE. YOUR SERVICE HAS RECENTLY SUFFERED SOME SETBACKS. I WARN THAT MR. BORIS YUZHIN (LINE PR, SF), MR. SERGEY MOTORIN,

(LINE PR, WASH.) AND MR. VALERIY MARTYNOV (LINE X, WASH.) HAVE
BEEN RECRUITED BY OUR "SPECIAL SERVICES."

Yuzhin, Motorin, and Martynov — Hanssen had named them all.
He knew, as he did when he named Polyakov, that they would likely
be dragged into the dread interrogation chambers of Lubyanka Prison
beneath Moscow Center. Again, he couldn't risk that one of them
might find out about him and tell the FBI.

Besides, he'd get paid faster. By sacrificing Yuzhin, Motorin, and
Martynov, chances were his credentials would be instantly established.
Even the most obtuse KGB bureaucrat would realize that no counter-
feit traitor would offer up the lives of three RIPs. Hanssen threw in a
few other items almost as valuable — some information about some
recent defectors, and something later referred to in legal documents as
"a particular highly sensitive and classified information collection
technique," most probably a reference to the hundred-million-dollar-
plus tunnel under the Soviet Embassy.

At this point, common sense suggests that the KGB officers in the
residence would have been eager to know the true identity of the
American who signed himself "B." Yet the KGB never learned who
Hanssen was. Why not? It would have been a simple matter to tail his
car. "The Russians could have spied on him had they chosen to," said
Tom Burns. "But to what purpose, and what were they risking if he
found out? He might terminate this coup. They had evaluated the
information. His identity would be interesting but not essential. They
would have had to do a risk-benefit analysis. They knew where the
drops were, but it could have exposed them."

On October 15, a parcel stuffed with classified documents arrived
at Degtyar's home. The next morning, an FBI surveillance team
posted in the "bucket," an office directly across Sixteenth Street NW
from the Soviet Embassy, dutifully noted that Degtyar showed up for
work with a big, black canvas bag instead of his usual briefcase. The
mood in the KGB residency must have been festive. The American
who signed himself "B" clearly could get his hands on some choice
goods. And he was delivering.

Nine days later, Degtyar received a letter with a New York post-mark. It contained instructions from "B" describing the Nottoway Park dead drop where the first installment of his money was to be deposited. Hanssen knew the location well: it was only two blocks from his home in Vienna, not a five-minute walk. His daughter, Lisa, was only thirteen days old, so he couldn't be gone for long.

DROP LOCATION

Please leave your package for me under the corner (nearest the street) of the wooden foot bridge located just west of the entrance to Nottoway Park. (ADC Northern Virginia Street Map, #14, D3)

PACKAGE PREPARATION

Use a green or brown plastic trash bag and trash to cover a water-proofed package.

SIGNAL LOCATION

Signal site will be the pictorial "pedestrian-crossing" signpost just west of the main Nottoway Park entrance on Old Courthouse Road. (The sign is the one nearest the bridge just mentioned.)

SIGNALS

My signal to you: One vertical mark of white adhesive tape meaning I am ready to receive your package.

Your signal to me: One horizontal mark of white adhesive tape meaning drop filled.

My signal to you: One vertical mark of white adhesive tape meaning I have received your package.

(Remove old tape before leaving signal.)

On November 2, KGB officers, following the directions, hid $50,000 in cash at the Nottoway Park dead drop. Six days later, Hanssen responded with an effusive letter. "I also appreciate your courage and perseverance in the face of generically reported bureaucratic obsta-cles," he wrote. "I would not have contacted you if it were not reported that you were held in esteem within your organization, an organiza-tion I have studied for years." He enclosed some TOP SECRET/SCI information about technical collections. And he renewed his efforts to get rid of Yuzhin, Motorin, and Martynov, writing:

> I cannot provide documentary substantiating evidence without arousing suspicion at this time. Never-the-less, it is from my own knowledge as a member of the community effort to capitalize on the information from which I speak. I have seen video tapes of debriefings and physically saw [Martynov], though we were not introduced. The names were provided to me as part of my duties as one of the few who needed to know. You have some avenues of inquiry. Substantial funds were provided in excess of what could have been skimmed from their agents. The active one has always (in the past) used a concealment device — a bag with bank notes sewn in the base during home leaves.

Certainly, the family needed every dime Hanssen could bring home. But Hanssen's letters to the KGB did not convey a sense of financial desperation. Anyone who had spent any time studying the KGB would know that the Soviets were notoriously cheap; anyone doing business with them was advised to drive a hard bargain. Bob Hanssen knew this better than most. Yet he was content to let the KGB bean counters decide what they wanted to pay. This strongly suggests a man who is not driven by money. Indeed, in his November 8 letter he referred to money with the airy nonchalance — distaste, almost — of a flower child with a trust fund. "As far as the funds are concerned," he wrote, "I have little need or utility for more than the 100,000. It merely provides a difficulty since I can not spend it, store it or invest it easily without tripping 'drug money' warning bells. Perhaps some diamonds as security to my children and some good will so that when the time comes, you will accept by [sic] senior services as a guest lecturer [sic]. Eventually, I would appreciate an escape plan. (Nothing lasts forever.)"

On November 6, 1985, Valery Martynov boarded an Aeroflot flight for Moscow. He assured his wife, Natalya, that the trip was routine, that he would be home in a matter of days. Ten days later, she got a note from Valery saying he was in the hospital with a knee injury from toting luggage. He wanted her and the children to join him in Moscow. Natalya obediently packed up and headed home on the next flight. Hanssen had gotten part of what he wanted: none of the FBI's Soviet RIPs remained in the States. The FBI might be able to get to one of them eventually, but as long as they were in Moscow, it wasn't going to be easy or soon.

Still, Hanssen was edgy, starting to jump at shadows. On March 3, 1986, a KGB officer filled the Nottoway Park drop with a payment, but Hanssen didn't show up to collect it. After a time, the Russians prudently reclaimed their cash. In a letter to Degtyar on June 30, 1986, Hanssen explained that he feared his cover may have been blown by KGB counterintelligence officer Viktor Gundarev, who had defected to the United States the previous February. To find out, he wrote, he had gotten hold of Gundarev's debriefing report. "I have only seen one item which has given me pause," he wrote. The FBI had asked if Gundarev knew Viktor Cherkashin. "I thought this unusual," Hanssen wrote. "I had seen no report indicating that Viktor Cherkashin was handling an important agent, and here-to-fore [sic] he was looked at with the usual lethargy awarded Line Chiefs. The question came to mind, are they somehow able to monitor funds, ie., to know that Viktor Cherkashin received a large amount of money for an agent? I am unaware of any such ability, but I might not know that type of source reporting."

The FBI would have loved to be half as good as Hanssen imagined, but in fact, the bureau had no pipeline into transfers of KGB funds between the KGB residence in Washington and Moscow Center, much less a line-by-line accounting of what every KGB officer was disbursing. But Hanssen's paranoia did not prevent him from, in the very same letter, dropping a devastating bit of information. Hanssen disclosed to the KGB that the United States was exploiting a specific technical vulnerability in Soviet communications satellite transmissions. With this casual statement, Hanssen wrecked the NSA's successful program, developed at a cost of untold millions of dollars, to intercept

Soviet government messages. Hanssen could have demanded any sum for this information, yet he merely dumped it out, almost as an after-thought.

Judging by the rest of Hanssen's June 30 letter, he was playing a game — for which he seemed to enjoy inventing elaborate rules. He wrote:

> If you wish to continue our discussions, please have someone run an advertisement in the Washington Times during the week of 1/12/87 or 1/19/87, for sale, "Dodge Diplomat, 1971, needs engine work, $1000." Give a phone number and time-of-day in the advertisement where I can call. I will call and leave a phone number where a recorded message can be left for me in one hour. I will say, "Hello, my name is Ramon. I am calling about the car you offered for sale in the Times." You will respond, "I'm sorry, but the man with the car is not here, can I get your number." The number will be in Area Code 212. I will not specify that Area Code on the line.

Hanssen was reaching back into his earliest police days for the code name "Ramon." "Ramon" had been the fictional name for an undercover agent invented by his instructor, Jack Clarke, during Hanssen's training for the Chicago Police Department's C-5 anticorruption unit, a guy whose best weapon had been his ability to blend into the gang he was infiltrating.

In an earlier letter, Hanssen had specified a time code. "I will add 6, (you subtract 6) from stated months, days and times." Following that code, from July 14 through July 18, 1986, the KGB placed an ad in the *Washington Times*. The ad contained a telephone number — 703-451-9780 — which rang at a public phone booth near the Old Keene Mill Shopping Center in Fairfax County, Virginia.

On July 21, 1986, Hanssen dialed the number. KGB officer Aleksandr Kirillovich Fefelov answered. The conversation was terse. Hanssen gave a phone number — 628-8047 — and hung up. Fefelov waited an hour, called the New York number, and said the Nottoway Park dead drop site was loaded. But there was a hitch: the package of cash wasn't where it was supposed to be, and Hanssen went home empty-handed. Bob upbraided the Soviets in an August 7 letter to Degtyar and said he would call the Old Keene Mill pay phone on August 18.

That day, Fefelov stationed himself by the phone and when the man he knew as "B" called, he taped the conversation, which went like this:

"B": Tomorrow morning?

FEFELOV: Uh, yeah, and the car is still available for you and as we have agreed last time, I prepared all the papers and I left them on the same table. You didn't find them because I put them in another corner of the table.

"B": I see.

FEFELOV: You shouldn't worry, everything is okay. The papers are with me now.

"B": Good.

FEFELOV: I believe under these circumstances, mmmm, it's not necessary to make any changes concerning the place and the time. Our company is reliable, and we are ready to give you a substantial discount which will be enclosed in the papers. Now, about the date of our meeting. I suggest that our meeting will be, will take place without delay on February thirteenth, one three, one p.m. Okay? February thirteenth.

Felefov gave Hanssen some more instructions about the pickup and how to notify the Soviets that Hanssen had received their payment before winding up the conversation.

FEFELOV: I hope you remember the address. Is . . . if everything is okay?

"B": I believe it should be fine and thank you very much.

FEFELOV: Heh-heh. Not at all. Not at all. Nice job. For both of us. Uh, have a nice evening, sir.

"B": Do svidaniya.

FEFELOV: Bye-bye.

Hanssen returned to the park at seven A.M. on August 19, 1986, and picked up a package that contained a new accommodations address, code-named NANCY, where he was to write. It was the Alexandria home of KGB Line PR officer Boris M. Malakhov, who replaced Degtyar when he rotated back to Moscow. There was an emergency communications plan so that Hanssen could contact the KGB residency in Vienna if he had to flee the U.S. There were proposals for two new dead-drop sites.

There was also $10,000 in cash. That, plus the money he had received on November 2, added up to $60,000. Hanssen had sold out NSA projects that had cost well more than $100 million, details on a few defectors, and four men's lives for a paltry sum. Moscow Center must have thought him either stupid or crazy.

He was neither. This wasn't just about money.

10

WILDERNESS

In October of 1986, the CIA station in Moscow reported a shattering event: KGB officers Valery Martynov and Sergei Motorin were on trial for treason. Martynov, who had accompanied Vitaliy Yurchenko home when he reneged on his defection, had been arrested at the Sheremetyevo Airport on November 7, 1985 — thirty-four days after Hanssen's letter informing on the young KGB officer had reached Viktor Cherkashin. Motorin had been arrested in November or December 1985. He had telephoned a woman friend in the States several months later and had said he was fine, but in retrospect, it seemed that his calls had been controlled by the KGB to mislead U.S. intelligence. Rumors had swirled for a while that the COURTSHIP recruits were in trouble. Now the worst had been confirmed.

Bill Mann, Martynov's recruiter, was at an in-service training session at the FBI Academy in Quantico, Virginia, when the agent sitting next to him whispered that Martynov was "missing in action and presumed dead." Mann had been transferred out of Washington in 1993 and had had no need to know about the operation: he and Valery were just friends. Mann felt sick and furious. He thought to himself that CIA turncoat Edward Lee Howard must have fingered Valery.

But if he hadn't? Howard hadn't been in a position to know everything. Martynov and Motorin might have made mistakes that gave themselves away. Or the FBI agents handling these men could have somehow stumbled on a KGB trip wire. Or perhaps the field office was penetrated by electronic means. "I thought they might be reading our codes," recalled an FBI Soviet section supervisor.

There was another, dreadful, possibility: another double agent. Perhaps more than one. But the evidence was too sketchy — nonexistent, really — to order up a full-fledged field investigation. Mole hunts are noisy affairs, not to be undertaken lightly. "If you shake the trees," said one agent, "the moles go to ground, and you never find them." Don Stuckey, the FBI Soviet section chief, and Bob Wade, his deputy, decided to create a small unit to run below the radar for a while. A half-dozen agents from headquarters and the Washington field office were tapped to pore over every piece of paper generated by Operation COURTSHIP, looking for anything that would shed light on the compromises of Martynov and Motorin. Tim Caruso, a headquarters Soviet section supervisor assigned to the group, named the unit ANLACE, for a double-edged medieval dagger, a traitor's weapon. Stuckey, Wade, and the few others who knew about the project called it, less grandly, "the guys in the vault," because the agents sat, day after day, in the space designated for the Intelligence Division: an ugly, windowless, triangular room with a combination lock, situated on the fifth floor of the Hoover Building.

In December 1986, Gus Hathaway, head of the counterintelligence staff for the CIA Directorate of Operations, the clandestine service, asked Stuckey, Wade, and the ANLACE FBI counterparts to meet at Camp Peary (aka "the farm"), a secret CIA training facility near Williamsburg, Virginia. Hathaway introduced Jeanne Vertefeuille, a Soviet/East European (SE) division hand who had been given charge of a new unit called, blandly, the Special Task Force, which would be staffed with one active and two retired SE veterans. The Vertefeuille group's mission was grim. The COURTSHIP recruits were just part of the picture. The CIA's entire Soviet intelligence program had been hemorrhaging like an ebola victim. Now there was scarcely any life left in it. The task force, set up in October, would analyze the losses and sort out which ones were truly Howard's doing and which had other causes, intentional or accidental.

The CIA officials went over the list of known and suspected losses and near misses. The first hint of trouble had come in late May of 1985, when Sergei Bokhan, a GRU officer recruited by the CIA and based in Athens, was recalled to Moscow. His American handlers were suspicious and advised him to defect, which he did. Then, on June 13, Moscow-based CIA officer Paul Stombaugh was ambushed and detained by the KGB on his way to meet Adolf Tolkachev, a Soviet expert on the U.S. Stealth aircraft program. Tolkachev's whereabouts were unknown until October 22, 1986, when Tass reported that he had been executed for high treason.

Edward Lee Howard might have known about Bokhan and Tolkachev, but there was only a slim chance he was aware of the next loss, Leonid Poleschuk, a KGB Line KR officer based in Lagos who had worked for the CIA years earlier in Katmandu but who had been dormant at the time Howard was fired. Poleschuk, reactivated in the spring of 1985, was considered a great catch because, as a Soviet counterintelligence officer, he would know about the Western double agents being run by Moscow Center. In May 1985, Poleschuk was recalled to Moscow and, within a few weeks, arrested.

KGB officer Gennady Varenik, based in Bonn, sent word of Poleschuk's arrest to Langley in late September or early October. Not long afterward, Varenik went into East Berlin and never returned. Also in October, Lisbon-based KGB officer Gennady Smetanin and his wife were recalled to Moscow and arrested. And while Howard might have known about Martynov, he probably had not known about Motorin, who was recruited later.

Vertefeuille theorized that the KGB had maintained secrecy as long as possible. "They had to roll up the ones who were overseas first, because if they started wrapping them up in the Soviet Union, the others would hear and they would defect," she said. "So they got them back as subtly as they could."

By the time Hathaway convened the CIA-FBI meeting, the losses, both in terms of arrests of agents and blown operations, were staggering. Two months later, Paul Redmond, head of the SE division's counterintelligence group, delivered a bleak conclusion to the CIA's front office: "It seems clear, if only from the statistics, that we have suffered very serious losses recently and that not all these compromises can be attributed to Howard."

The FBI officials briefed the CIA managers on their own problems, offered leads, and consulted occasionally with Vertefeuille's team. But each agency's team worked separately; neither opened its files to the other until 1991. CIA officials, even more skittish about secrecy than their FBI counterparts, were determined to keep the Vertefeuille inquiry subterranean. How could the United States hope to attract new sources if word got around that the mortality rate among the old ones was just about 100 percent?

Besides, counterespionage inquiries were about as popular at Langley as temperance campaigns. Everyone still mourned the wrecked careers and missed opportunities caused by James Jesus Angleton, who ruled over the agency's counterintelligence program from 1954 to 1974 and whose suspicions ran so rampant that at one point, acting on a tip, he investigated every CIA employee whose name began with "K." Angleton was right, in principle. There had been double agents — Barnett, Miller, Howard, Pelton, Pollard, Wu-Tai Chin, and the Walker-Whitworth ring, to name a recent few — and there would be more. But Angleton's methods were all wrong. They wrapped the agency's Soviet program around an axle, ruined many innocent people's lives, and perversely helped the Soviet Union, by alerting those who were guilty.

Still, there seemed to be a good chance that there was a penetration somewhere, either human or electronic. As early as January 1986, SE chief Burton Gerber had set up a "back room" operation to limit distribution of information about new recruits to himself, Paul Redmond, and a few others. Sandy Grimes, an intelligence officer who had handled Dmitri Polyakov and Leonid Poleschuk, was put in charge of the "back room," and it seemed to be working; the Soviet assets recruited after 1985 were still alive.

The trek through years of files was slow, painstaking, and frequently sidetracked by false leads. Moscow Center took advantage of the disarray in Langley by generating disinformation that sent the Vertefeuille task force off on fool's errands, planting rumors that Howard was responsible for more compromises than had been realized, that the agents themselves or their CIA handlers had made fatal errors in their tradecraft, or that an electronic penetration was to blame. The phrase "wilderness of mirrors," an allusion to a line from T. S. Eliot's poem "Gerontion," which James Angleton had used to describe the world of counterespionage, had never seemed more apt.

But while the vault guys and the Vertefeuille group succeeded in keeping their work secret, they didn't solve any of the mysteries they were supposed to solve. In a 1994 postmortem, then–CIA inspector General Frederick Hitz found that "prior to 1991, no formal lists of suspects based on access were created or reviewed [by the Vertefeuille team]. The SE Division had not even maintained accurate 'bigot' lists of people who had had access to the blown operations." Vertefeuille replied that her team had compiled long lists of people and units that had known about the compromised cases, but they were never complete, because the access records left a lot of people out. Certainly they never reduced the suspects to a short list.*

---

*To make matters worse, in December 1986, CIA director William Casey, a driving force in the search for the truth behind the 1985–86 compromises, suffered a seizure caused by a brain tumor and lapsed into an irreversible coma. Casey's deputy, Bob Gates, had his hands full with the Iran-Contra scandal, which broke in October 1986, when an illegal CIA arms pipeline to the Nicaraguan guerrillas was exposed. By early 1987, the scandal was engulfing much of the agency's Directorate of Operations, the clandestine service. Casey finally died on May 6, 1987, and twenty days later, FBI director William Webster was sworn in to succeed him. Webster immediately became preoccupied with assembling a staff and putting out brushfires.

Over at the FBI, the ANLACE agents were also stuck. They were good investigators, but they weren't investigating. They were limited to reading old paperwork, and the answers just weren't there. Without the license to go out, develop leads, squeeze people for information and break a few dishes, they couldn't advance the case. And their task was harder than the CIA's because there were too many people who knew too much. The ANLACE team didn't even bother to develop a list of suspects to be scrutinized for the usual vulnerabilities — booze, broads, and money, in bureau shorthand — realizing that 250 FBI personnel in the Washington field office *alone* knew about Martynov and Motorin, not to mention many more at FBI headquarters (as well as, obviously, the CIA).

The ANLACE inquiry wound up in September 1987, making no conclusions about how COURTSHIP had been blown. By that time, Motorin had been shot by a firing squad (in February 1987), Martynov had been executed (in May), and Boris Yuzhin had been convicted of high treason and sentenced to fifteen years in the Gulag.

Bob Hanssen had gotten his way. There was nobody left inside the KGB to rat him out.

On August 3, 1987, Hanssen was transferred from New York back to Washington, where he resumed his old job as a supervisory special agent in the Soviet analytical unit. The lateral move signaled that Hanssen's tour as a technical squad supervisor in New York had been a flop. If he had done well, the natural progression would have been for him to return to headquarters as a grade 15 unit chief, then go out to the field as an assistant special agent in charge. The best — or at least the best-connected — middle managers ascended to the rarefied ranks of the Senior Executive Service, becoming section chiefs at headquarters or, glory of glories, special agent in charge of one of the fifty-six field offices. To stall at the rank of supervisor — this was a vote of no confidence, and everybody knew why: "He had no interpersonal skills," said one bureau official who knew Hanssen in those days. "He could write and analyze well, but he was a loner, not the personality to be a field supervisor."

While the move to Washington was not a career booster, it was welcome news to Hanssen's family. Among other things, the Virginia suburbs were not nearly as expensive as those in Westchester County. Bob and Bonnie bought a four-bedroom ranch house on Talisman Drive, several miles from where they had lived before. They enrolled Jane and Susan, already teenagers, in Oakcrest, then located in a wealthy residential section of Washington, D.C. Although the Oakcrest tuition was $4,900 per student, the school offered scholarships and tuition breaks to many large Catholic families like the Hanssens, based not only on need but also on the amount of time parents were willing to donate to school projects. Bob and Bonnie were quick to volunteer. "As a parent team, they were singular," said Oakcrest director Barbara Falk. "They have always been extremely supportive of whatever we needed, from painting to computer work." The two older boys, Jack, eleven, and Mark, seven, attended the parochial school affiliated with Our Lady of Good Counsel, where Hanssen attended daily mass. The two youngest children, Greg, four, and Lisa, two, were not yet in school.

Hanssen had not communicated with the KGB while the ANLACE project was going on. But on September 8, 1987, about the time the vault guys were wrapping up their unsatisfying final report, he wrote KGB Line PR officer Boris M. Malakhov, his new KGB contact, using the alias and the return address: R. Garcia, 125 Main St., Alexandria, Va. Hanssen picked up exactly where he had left off, with "Ramon" briskly rejecting several dead-drop locations proposed by the KGB the year before:

Dear Friends:

No, I have decided. It must be on my original terms or not at all. I will not meet abroad or here. I will not maintain lists of sites or modified equipment. I will help you when I can, and in time we will develop methods of efficient communication. Unless a [sic] see an abort signal on our post from you by 3/16, I will mail my contact a valuable package timed to arrive on 3/18. I will await your signal and package to be in place before 1:00 pm on 3/22 or alternately the following three weeks, same day and time. If my terms are unacceptable then place no signals and withdraw my contact. Excellent work by him has ensured this channel is secure for now. My regards to him and to the professional way you have handled this matter.

Sincerely,

Ramon

Three days later, on September 14, Malakhov received a package of documents, some of which were stamped TOP SECRET and had originated at the National Security Council. The day after that, KGB agents filled the dead drop in Nottoway Park with $10,000 and a letter proposing two new dead-drop sites in suburban Virginia. One, code-named AN,* was in Ellanor C. Lawrence Park, south of Dulles Airport, and another, code-named DEN, was even farther away. Hanssen would have none of it, responding in a note left in the Nottoway drop site:

My Friends:

Thank you for the $10,000.

I am not a young man, and the commitments on my time prevent using distant drops such as you suggest. I know in this I am moving you out of your set modes of doing business, but my experience tells me the [sic] we can be actually more secure in easier modes. . . . I will clear this once [at] 'AN' on your scheduled date (rather than the other). . . . Find a comfortable Vienna, Va. signal site to call me to an exchange any following Monday.

Good luck with your work.

Ramon

On November 10, 1987, Malakhov received a letter with the return address, "J. Baker, Chicago." Chicago, of course, was Hanssen's home-

*The Soviets named the dead drops alphabetically.

town, and Hanssen may have been making a tongue-in-cheek reference to James A. Baker III, President Reagan's secretary of state. In his letter, Hanssen said he planned an "urgent" delivery for Monday, November 16. The KGB filled the AN dead drop with a package of cash, but Hanssen did not claim it on the sixteenth. The next day, using the return address "G. Robertson, Houston," he wrote a testy letter suggesting that he preferred the asphalt paths of nearby Nottoway Park to unknown spots in the Virginia countryside:

> Unable to locate AN based on your description at night. Recognize that I am dressed in business suit and can not slog around in inch deep mud. I suggest we use once again original site [Nottoway Park]. I will place my urgent material there at next AN times. Replace it with your package. I will select some few sites good for me and pass them to you. Please give new constant conditions of recontact as address to write. Will not put substantive material through it. Only instructions as usual format.
>
> Ramon

Hanssen's November 23 delivery to the Nottoway Park drop site must have delighted the Russians. It included highlights of the debriefings of defector Vitaliy Yurchenko as well as technical data on COINS-II (Community On-Line Intelligence System), the classified intranet used by the U.S. intelligence community. This information could be immensely helpful to KGB hackers.

The KGB stuffed the drop with $20,000 in cash and a grateful letter from KGB director Vladimir Kryuchkov, the tough, ambitious protégé of the late Soviet leader and former KGB head Yuri Andropov. An able manipulator, Kryuchkov stroked Hanssen with the soothing message that $100,000 had been deposited for him in a Moscow bank. Hanssen knew the promise was meaningless, but he liked being appreciated.

The KGB package also held two addresses of Soviet diplomats to be used for "B's" communications, and also two new proposals for dead-drop sites. Hanssen sent a one-word note in response: "OK." He used the return address "Jim Baker, Langley." Hanssen, playing mind games again, probably knew the Soviets would be amused at his reference to the address of CIA headquarters. He was spying, literally, in the agency's backyard. Hanssen, himself, was undoubtedly amused, too.

In December 1986, a U.S. Marine named Clayton Lonetree had informed a CIA officer in Vienna that while posted as a guard at the U.S. Embassy in Moscow in 1985 and part of 1986, he had an affair with a Russian woman working for the KGB and had given a KGB officer, among other items, an embassy phone book. Another Marine, Arnold Bracey, embellished the story, claiming that Lonetree had let KGB agents into the embassy. Bracey eventually recanted this statement, and Lonetree passed a polygraph denying Bracey's allegations, but not before the incident caused embarrassing headlines, and not before the lives of hundreds of Moscow-based U.S. foreign-service officers and their families were severely disrupted.

At a meeting in February 1988, the Vertefeuille group told FBI intelligence officials they had discounted Lonetree as a possible source of the 1985–86 compromises. They were working on two other leads. One involved a CIA SE Division employee who had problems during routine polygraphs and was spending more money than his salary would allow. Another, from an anonymous tipster, was that the CIA communications center in rural Warrenton, Virginia, had been penetrated electronically and that an employee there had been compromised. But the tip on the Warrenton center was eventually thought to be a KGB plant, and the SE Division employee's spending spree was explained by his wife's recent inheritance. Both investigative efforts ate up precious months.

All of which was good news for several moles, including Bob Hanssen. On February 8, he filled the Nottoway Park drop with documents and a computer diskette naming a KGB "illegal," who had become a RIP for the FBI, even passing on the man's KGB code name and recent assignment. There was a description of the latest U.S. technology for intercepting the communications of the Soviet Union and other governments of interest to the United States. In addition,

Hanssen gave the Soviets information on "a particular limitation of the NSA's ability to read certain Soviet communications." This insight was invaluable to the Soviets as they strove to elude the NSA's global array of antennae.

Most important, he told the KGB all about his new friend, Viktor Sheymov. Viktor Sheymov was one of the most important defectors ever to arrive in the West. He was a promising young mathematician and engineer in 1971, and the KGB, recognizing his brilliance, tapped him for its Eighth Chief Directorate, the Soviet equivalent of the National Security Agency. Sheymov soon displayed his proficiency for improving the essential tools of all spymasters, the science of covert communications, encryption, and decryption. By 1980, he was the youngest major in KGB history and in charge of security for all KGB overseas communications. His team created codes and then protected them so that Soviet embassies, military leaders, and intelligence officers could communicate with one another across vast distances. The information he had was considered so valuable that when he traveled to Bloc countries, he was closely watched by KGB minders. He could never even stay in a hotel: in whatever city he visited, he was required to sleep in the Soviet Embassy.

Like many Russian intellectuals, Sheymov loved his country but despised the Soviet system and railed at being forced to serve the hated KGB. "It was the most horrible political system on the face of the planet," he said. "The Communists killed many more people than anybody else. They smashed every dissident group. My objective became to inflict as much damage as I could to the Communist Party and the KGB. The most effective way was to go to the United States."

Sheymov figured out a way to contact a CIA officer while on a trip to an Eastern European capital. He was ready to defect, he said, but only if he could bring his wife, Olga, an artist, and their four-year-old daughter, Elena. CIA officers saw instantly how much Sheymov's knowledge would be worth to the West, especially if they could somehow keep his defection a secret so that the Soviets did not change their communications systems and codes. Over the next three or four months, the CIA station in Moscow made elaborate plans for a "black exfiltration," the first the agency had ever attempted. This was an exquisitely choreographed maneuver in which the family simply disappeared without a trace — no good-byes to relatives, no packing. The technique required split-second timing and coordination among CIA stations all along the escape route to the West. Until now, there had never been a prospective defector who was worth all that trouble. "Timing is everything," explained David Rolph, a Moscow-based CIA officer who met Sheymov in back alleys to work out the details. "If he'd come along a few years later, we might not have been so interested, but at the time, he had very high-level access to extremely critical information that we really needed, so we were willing to mount what became a very complex and dangerous operation."

One morning in May of 1980, Viktor and Olga got up, dressed Elena, made coffee and breakfast, opened the newspapers — then walked out the door and never looked back until they arrived in the West. Sheymov's KGB minders, when they arrived, might have imagined — as the CIA hoped they would — that the family had been mugged and murdered in a park, and their bodies tossed in a lake.

The family was relocated in the Washington area under a new identity, and Viktor worked as a consultant to the NSA, providing the missing pieces that permitted U.S. intelligence to break into the KGB and Soviet military communications he himself had helped encode.

Then Sheymov met Bob Hanssen. They were introduced in early 1988, when Hanssen became Sheymov's liaison to the FBI. Before long, Viktor and Olga were having family dinners at the Hanssen household with Bonnie and all the kids. Sheymov, a cosmopolitan who had a quick, thoroughly secular wit and bore a slight resemblance to the German actor Oscar Werner, seemed to have little in common with the pious, parochial FBI agent. Under ordinary circumstances, they most likely would never have become friends. But Sheymov's situation was anything but ordinary. As a fugitive forced to reinvent his entire life, he needed all the help he could get. Once he got to the United States, he found CIA headquarters worse than useless. By contrast, Hanssen's square competence seemed comforting.

"He was a good man, very religious, very conservative," Sheymov said. He came to see Hanssen as a big brother, who could guide him through the Washington maze. Hanssen, in turn, behaved very protectively toward Sheymov, refusing to let other FBI agents talk with him unless he, Hanssen, was in the room.

Hanssen later bragged to a friend that he had saved Sheymov's life when a KGB "wet affairs" team — a hit squad — was gunning for him. Sheymov, when asked about this claim after Hanssen was arrested, dismissed the story as one more product of Hanssen's overactive imagination. Sheymov recalled talking to Hanssen about his worries that the KGB would find out where he was and try to kill him. But Sheymov said he never faced any actual attempt on his life, and Hanssen certainly never dashed in to hold off Soviet assassins.

Hanssen's February 8 report was the first of many messages he would send the KGB about Sheymov's work at the NSA. It was not until Hanssen's arrest that Sheymov realized the crushing truth. Hanssen had been cynically cultivating him and his wife, socially as well as professionally, while quietly reporting their activities back to the KGB. Hanssen was selling out Sheymov to the very agency he and his wife had risked their lives, and their daughter's life, to destroy.

Not surprisingly, the KGB was grateful. The Soviet spy-catchers filled the drop with another token of appreciation: $25,000, and a letter of thanks from Kryuchkov that included a request for more detail about the FBI's agent network in New York City.

On March 16, 1988, Hanssen's second computer diskette arrived at a KGB "accommodation address," the home of a minor Soviet diplomat who was serving as a go-between for Hanssen's mail. The return address was again "Jim Baker, Chicago." Another diskette came by mail ten days later.

On March 28, the KGB deposited $25,000 in the dead drop, along with a request for strategic information on codes, U.S. submarine movements, and the Strategic Defense Initiative, the hugely expensive, and mostly theoretical, plan for a missile shield. The note added that the diskettes left earlier were blank. Still using the "Jim Baker" alias,

Hanssen fired back an impatient note bristling with computer-geek exasperation. "Use 40 TRACK MODE." Translated into laymen's terms, this meant that the data had been hidden on certain tracks that could be found only with software employing the same codes. Someone who found the disk and tried to read it the normal way would think it was blank.

A few days later, Hanssen sent a diskette with extensive information about a KGB officer who was being studied by the FBI, two more Russians recruited by the FBI, and more on Sheymov. For the Soviets, Bob was the sort of spy they had dreamed about. The goods just kept coming, and coming, and coming.

Hanssen, meanwhile, was using his computer skills to ingratiate himself with "seventh floor," bureau-speak for the suites occupied by the director and his senior aides. Hanssen often volunteered to do the sort of computer work that other agents disliked. In early 1988, the seventh floor was in a swivet over the CISPES affair. This was an overly long, overly zealous Intelligence Division probe of the leftist Committee in Solidarity with the People of El Salvador, which an FBI informant had wrongly claimed was plotting terrorist acts. The division had run an undercover informant into CISPES and kept the case open long after it was clear the group was not involved in criminal activity. This had violated the Justice Department's post–Church Committee guidelines prohibiting FBI monitoring of peaceful dissident groups. Director William Sessions, the former federal judge from Texas who had succeeded Webster, and Executive Assistant Director Oliver B. "Buck" Revell had to brief Congress about what the FBI had done, and when, in the CISPES case. This was not as easy as it seemed: the bureau's computer software was capricious, and it could not be trusted to find all the records on any given topic.*

*Even in 2001, the FBI's computer and communications technology was eight to ten years behind private industry. The aging computers in most field offices could not even access the Internet, a tool available to most of the nation's schoolchildren. Moreover, in the wake of the September 11 terrorist attacks, it was disclosed that for some years, the FBI had so few Arabic linguists that it had not been able to read piles of documents seized during investigations of international terrorist cells.

Hanssen came up with some programming language that seemed to be an improvement in the way records were stored and retrieved. "It sounded good to me," said David Keyes, an Intelligence Division careerist who was Revell's special assistant during the CISPES brouhaha. "I thought he was a very bright guy. He had a lot of good ideas. He worked very hard and from that standpoint was really a positive example."

Hanssen also impressed the analysts who worked for him. "He was a solid citizen, a person I very much admired," said analyst Paul Moore, who sometimes carpooled with Hanssen. "He was very, very intelligent and capable, very much into family values, trying to lead a good life according to certain principles. I would like to have been more like a Bob Hanssen."

Moore especially admired Hanssen's ease with conceptual thinking and abstractions. Bob was good at making connections between two or three different events, and seeing patterns that would continue to play out. "Your job as an analyst is to explain how the present relates to the past and how it relates to the future," said Moore. Moore was struck, particularly, by a typical Hanssen saying: "Policies are constraints; constraints breed patterns; patterns are noticed." Bob had repeated these lines when teaching analysts. It would also appear in one of his letters to the Russians, a near giveaway clue had anyone like Moore seen the missives.

Hanssen disguised his treachery in a seeming dedication to improving the bureau's analytic product. "He wasn't someone who was maniacally trying to hurt the U.S. at every opportunity," Moore said. "Maybe after dark, but during the day he was going beyond what he had to do to help the country." At the office, Moore saw only Good Bob, the serious, religious, patriotic family man. Only later, when it was too late, did Moore realize his friend had been "howling at the moon at night." "Plainly, he had some compartments in his life," Moore said. "He just opened a door and went into Bad Bob's compartment, and there, he was very bad." And as bad as Bob had been, things were only going to get worse.

11

RAMON GARCIA

Spying under an alias is a lot like carrying on an anonymous affair with a stranger. When you are the unattainable object of desire, you hold all the power in the relationship. You don't have to compromise, be reasonable, see another's side, share, cope. You don't even have to be civil. You can walk out anytime you like. For a few hours, the door is shut on the clamor of daily life, on the banal obligations to family, work, church, and community. It's a walk on the dark side of the moon, not a place for the fainthearted, but if you can stand the cold and the loneliness, you're free. No wonder Robert Hanssen found it intoxicating.

The KGB officer was three minutes late. It was 9:03 P.M. on May 30, 1988, and Bob Hanssen was seething in the dark parking lot. Bob, using his old "Ramon Garcia" alias, had given the KGB strict instructions to leave the cash at Nottoway Park before nine P.M. But it wasn't there. Maybe the officer had gotten lost, or his wife sent him to pick up a child at the last minute, or he'd had car trouble. Whatever. This was not "Ramon's" problem. "Ramon" was a star. Stars don't have to understand.

Hanssen removed the vertical strip of white adhesive tape from the park sign, his signal that he was ready to receive a drop, and drove off, just as the Russian was pulling into the parking lot. Let the poor sap wonder what Moscow Center would do to him when the bosses found out their officer had caused the most brilliant double agent in the history of the Cold War to quit the game.

But Bob Hanssen was having too much fun playing "Ramon" to walk away. On June 13, he wrote a letter venting a bit of pique, but also bestowing forgiveness and conveying a knowing professionalism as frosty as one of James Bond's martinis.

I found the site empty. Possibly I had the time wrong. I work from memory. My recollection was for you to fill before 1:00 a.m. I believe Viktor Degtyar was in the church driveway off Rt. 123, but I did not know how he would react to an approach. My schedule was tight to make this at all. Because of my work, I had to synchronize explanations and flights while not leaving a pattern of absence or travel that could later be correlated with communication times. This is difficult and expensive.

I will call the number you gave me on 2/24, 2/26 or 2/28 at 1:00 a.m., EDST. Please plan filled signals. Empty sites bother me. I like to know before I commit myself as I'm sure you do also. Let's not use the original site so early at least until the seasons change. Some type of call-out signal to you when I have a package or when I can receive one would be useful. Also, please be specific about dates, e.g., 2/24. Scheduling is not simple for me because of frequent travel and wife. Any ambiguity multiplies the problems.

My security concerns may seem excessive. I believe experience has shown them to be necessary. I am much safer if you know little about me. Neither of us are children about these things. Over time, I can cut your losses rather than become one.

Ramon

P.S. Your "thank you" was deeply appreciated.

Spies, like femmes fatales, have to keep their mojos working or the game gets stale. Every time the action started to sag and "Ramon" felt that he was being taken for granted, Hanssen spiced things up with some information guaranteed to make his handlers sit up straight. On July 18, he filled the Nottoway Park drop with 530 pages of pure gold. Among the nuggets:

- An intelligence analysis discussing how much data Soviet intelligence had collected about U.S. nuclear weapons capabilities. This document discussed the U.S. early warning systems and systems to defend or retaliate against a large-scale nuclear attack.
- A report about U.S. nuclear programs dated November 1987.
- A report of the Intelligence Producers Council, an arm of the Defense Intelligence Agency, titled "Compendium of Future Intelligence Requirements: Volume II." This included a "comprehensive listing" of what the U.S. intelligence community had been tasked to find out about the military capabilities and preparedness of the Soviet Union and other nations.
- A document describing the NSA's inability to read certain Soviet communications.
- A CIA counterintelligence staff study titled "The Soviet Counterintelligence Offensive: KGB Recruitment Operations Against CIA," dated March 1988.
- A comprehensive historical FBI review of allegations from various RIPs and defectors about supposed penetrations of the U.S. intelligence community. This document described the FBI's sources, not by true name but with enough detail to permit a savvy KGB counterintelligence team to triangulate on some or all of them. It outlined what each source had said in detail, and what the FBI had done to investigate each allegation. This document was breathtaking in its scope, detail, and in the insights it offered. It gave the KGB an extensive set of facts about the U.S. intelligence community's recruitment program and, by explaining which aspects of which operations certain FBI officials regarded as having gone well or badly, a snapshot of how the FBI thought.

The KGB replaced the documents with a payment of $25,000. If Hanssen had forced the Soviets to bid for them, he could have put several new roofs on his house. He might even have gone home with enough money to retire. One of the many paradoxes of Bob Hanssen's psyche was that he placed a low material value on his work, even as he held himself to be vastly superior to everyone with whom he worked. It rankled Hanssen endlessly that most of the people vaulting past him up the FBI ladder weren't nearly as well read as he, nor as insightful, nor as well prepared for the blossoming of the information age. He didn't suffer fools gladly, and those he took for fools felt the same way about him. In a meeting, when colleagues offered different perspectives or challenged him, he bridled. "You could see him lock down," said Harry "Skip" Brandon, who became deputy assistant director of the Intelligence Division in 1980. "Most of us would encourage some back and forth, an exchange of ideas. But he was very rigid. As a result, he wasn't listened to as much as others." The KGB, which truly appreciated his services, would have showered him with spectacular rewards if he had insisted, yet over and over again, he let the buyers set their own price, and they low-balled him.

Perhaps he felt that negotiating would mark him as a common money-grubber like crude, greedy John Walker, sodden Edward Lee Howard, or Richard Miller, who was pathetic. Did he rationalize his crimes by telling himself he was simply playing intellectual tricks on a system that didn't deserve respect? Or did his reluctance to make monetary demands have something to do with the mixed messages he had received from his father? Howard Hanssen had always thought Bob was too smart to be a cop. At the same time, though, Howard had usually had something cutting to say about whatever Bob did achieve. Maybe, despite everything, Bob hadn't been able to shake the feeling that he didn't really deserve very much, even though most everyone in the family said that Howard was just bitter and mean. Bob's low opinion of the FBI may have had something in common with Groucho Marx's famous rejection of a snooty social club's invitation to membership. Sniped Groucho, "I don't care to belong to any club that will accept me as a member."

Though many of his FBI colleagues were long-standing Roman Catholics, none professed to be as devout as the convert Hanssen. Many of them drank, told lewd jokes, swore like stevedores, and went to mass only when their wives dragged them along. Around five P.M., many FBI headquarters guys pulled out of the FBI garage and straight across the street to Karpel's liquor store, which they called Carpool's. The driver would double-park while somebody would dash in for a six-pack to mellow out the boring drive home to the Virginia suburbs. Hanssen was never among them. A joke around the Intelligence Division was that if the Russians or Chinese really wanted to know what was going on, they should bug the tables in Hammill's back room. But Bob would never be seen at the smoky steak-and-fries restaurant that FBI executives and Justice Department lawyers regarded as their luncheon club. Nor did Hanssen attend the occasional bachelor parties at strip joints like Joanna's on M Street or Good Guys on Wisconsin Avenue. He also had very strong opinions about extramarital affairs, opinions he was not afraid to share. He told his friend Paul Moore, "If you went out to cheat on your wife, this would be a sinful thing to do, and also sinful for the woman you were cheating with. You were helping her to commit a sin." Added Moore, "He was rock-solid on these things."

Hanssen's transactions with the KGB laid waste to a great deal of the life's work of those who offended Bob. The destruction was especially severe in the New York office, the scene of Hanssen's most public humiliations. "Nothing would give him a greater kick than watching us running around in circles, while he was the puppetmaster, knowing all our efforts were in vain," said Jerry Doyle, who had been among those who rebuffed Hanssen's probing questions when they served in New York together. Jim Burnett, who worked on New York counterintelligence investigations during both of Hanssen's tours in the city, said, "I know several agents who spent a decade or more doing extraordinarily significant work and saw the results of it obliterated by what Hanssen betrayed." For instance, on September 21, Hanssen/Ramon sold the KGB information about FBI efforts to recruit certain Soviet diplomats, and on September 26 he passed along a memo about a New York–based KGB officer the bureau would like to recruit, an assessment of Soviet electronic surveillance systems, and the transcript of a CIA Counterintelligence Group meeting.

Hanssen's payment for all this was a diamond worth about $25,000, a good-size rock that the KGB left in a dead drop it coincidentally called "BOB," a footbridge in Idylwood Park, between Vienna and Falls Church. There was a note of thanks from KGB director Kryuchov, an assurance that an additional $50,000 had been deposited in his name in an account in Moscow, and a wish list asking for details of U.S. and Western human and technical sources, including U.S. electronic intelligence collection systems deployed covertly within the Soviet Union.

The day after Christmas 1988, Hanssen broke away from his family and drove to "CHARLIE" dead drop, the footbridge in Eakin Park, south of Vienna. There, he deposited 356 pages, including a report about U.S. targeting of certain Soviet communications, a document titled "Soviet Armed Forces and Capabilities for Conducting Strategic Nuclear War Until the End of the 1990s." The KGB responded with $10,000 in cash and a second, smaller, diamond worth approximately $18,000.

On March 20, 1989, Hanssen went back to "CHARLIE" with a diskette, about 539 pages of documents, and a report called "DCI Guidance for the National MASINT Intelligence Program (FY 1991–FY 2000)." MASINT stands for Measurement and Signature Intelligence and involves cutting-edge technologies that may someday be used to "sniff" chemicals and nuclear materials from a great distance. Eventually, CIA officials hope this technology will help the U.S. to determine whether chemical, biological, or nuclear weapons are being produced inside a camouflaged facility. The MASINT report was so hard to get that Hanssen asked the KGB to return it after it had been read and copied. It was yet another bonanza for Hanssen's handlers: a highly secret consensus within the U.S. intelligence community on the specific objectives of the program and the specific studies needed to develop those capabilities. In return, the KGB left $18,000 cash and a third diamond, worth about $12,000.

While Hanssen was busily pitching secret documents to the KGB, the CIA and FBI mole hunts were open for business in name only. Neither agency was any closer to explaining how the KGB had been able to roll up most of the U.S. spy network operating inside the Soviet system in 1985 and 1986. The Vertefeuille team believed that sixteen Soviet agents had been compromised, that Edward Lee Howard had certainly known of three of them and that he might have known of seven more. That left six that Howard could not have betrayed. However, no ill had befallen any of the Soviets recruited by the CIA *after* 1985. Did that mean that the mole, if indeed there had been one, had retired, been transferred, gone dormant, or died? Was someone still spying but thwarted by the "backroom" security measures from finding out about new recruits?

Nobody in senior or middle management ranks at either agency seemed to care enough to find out. The inquiries were simply forgotten, because of malaise, neglect, and distraction. At the CIA, nearly every executive involved in launching the investigation had been caught up in the Iran-Contra scandal. In the chaos that followed, their successors were diverted by their own struggles to survive. At the FBI, an ever-changing roster of senior and mid-level executives didn't even consider diverting resources to solving problems that had occurred on someone else's watch.

The FBI's lack of focus served Hanssen well. It was also good for the careers of the KGB officers who were handling "B"/"Ramon Garcia." In April 1989, several of them were decorated with the Order of the Red Banner, the Order of the Red Star, and the Medal for Excellent Service.

On May 22, Hanssen returned two of the diamonds and asked for money instead. Diamonds, after all, might draw attention to him if he tried to unload them. He wrote the KGB saying he wanted bonds and a bank account in Switzerland. He also told the KGB about a new FBI double-agent investigation of a U.S. State Department diplomat named Felix Bloch. Bloch had been under suspicion for some time, but what really tipped the scales was an April 27, 1989, telephone call, intercepted by the CIA between Bloch in Washington and Reino Gikman, a Vienna-based KGB "illegal." What was said between the two men didn't amount to an airtight espionage case, but it was enough for the FBI to place Bloch under surveillance. The CIA joined in. Its Paris station reported Bloch was seen meeting Gikman there on May 14. CIA officers in Brussels observed a Bloch-Gikman rendezvous on May 28.

Less than a month later, on June 22, as Bloch was preparing to meet with Gikman for a third time, an FBI wiretap on Bloch's line picked up a call from a man who introduced himself as Ferdinand Paul. Paul said that he was calling for "Pierre," whom the FBI later identified as Gikman. "He cannot see you in the near future," Paul said. "He is sick. A contagious disease is suspected." "I am worried about you," Paul added. "You have to take care of yourself."

The FBI agents who monitored the call were sure Paul was warning Bloch that he was being investigated. In a last-ditch effort to salvage the case investigation, agents confronted Bloch, hoping to jar him into confessing. But Bloch calmly denied any wrongdoing. Gikman, meanwhile, had left Vienna for Moscow. The FBI put a tight tail on Bloch, hoping to pressure him into breaking down and blurting out something useful. Not surprisingly, these heavy-handed tactics attracted unwanted attention. Before long, the network news broadcasts featured footage of a smirking Bloch trailed by serious young men with short haircuts and dark suits. The FBI looked ridiculous. Bloch was forced out of the State Department but would never be prosecuted.

Felix Bloch's name hung over the Hoover Building like a bloody shirt. Nobody in the Intelligence Division would forget this humiliation. FBI counterintelligence hands immediately suspected that somebody at State was to blame. Everybody there who had any inkling of the suspicions about Bloch was polygraphed, and everyone passed. After that, whenever information about a possible human penetration of the U.S. government arrived in the FBI Intelligence

Division, one of the first questions was: "Could he have tipped the KGB about Bloch?" If that was in the realm of possibility, the next question would be: "Was he Ferdinand Paul?"

Bob Hanssen was perhaps the only person on the Hoover Building's fifth floor who wasn't obsessed with the Bloch case. He didn't have to be. He knew all about Ferdinand Paul. Let the rest of them chase shadows — he *was* Ferdinand Paul.

On August 7, Hanssen filled the CHARLIE dead drop with five rolls of film containing photos of a highly restricted TOP SECRET/SCI analysis of the Soviet intelligence threat to what the Justice Department would later describe as a "highly compartmented US government program to ensure the continuity of government in the event of a Soviet nuclear attack, which analysis directly concerned means of defense or retaliation against large scale nuclear attack and other elements of defense strategy." It was a report that gave Soviets an intimate look into the way American officials thought about the unthinkable, a prize that would be talked about in Kremlin circles for years to come: the U.S. government's plans regarding what it would do if nuclear war broke out. Hanssen had delivered one of the crown jewels.

The KGB paid Hanssen $30,000 cash in the drop and a promise of more when the diamonds were liquidated. It proposed a new dead drop, ELLIS, in Foxstone Park, only a mile from Hanssen's house on Talisman Drive. It turned down his request for a Swiss account, without explanation.

On September 25, at a drop site called DORIS, Canterbury Park in Springfield, Virginia, Hanssen delivered another extremely valuable document, classified TOP SECRET/SCI. It contained details of an electronic penetration of a Soviet facility and a project aimed at intercepting Soviet communications overseas. The KGB paid just $30,000, an infinitesimal fraction of the program's cost to U.S. taxpayers.

On October 23, Hanssen initiated the new ELLIS dead drop, delivering information about a recently developed NSA capability to read certain Soviet communications. The KGB responded with $55,000 cash and a letter noting a $50,000 deposit to his account in Moscow.

After the wad of cash sat under the bridge across Wolftrap Creek for three days, the KGB retrieved it. Hanssen did not explain why he had not picked up the money, and on the last day of the month the Russians tried again. (Trick-or-treaters roaming the neighborhood on Halloween night might have thought the guy in the trenchcoat wandering off into the park was just dressed up as a spook, but in fact he was a real KGB officer.) The package also contained a KGB diskette with further instructions and a new mailing address. This time, Hanssen retrieved it.

In November 1989, CIA SE Division intelligence officer Diane Worthen, who had been working with her colleague I/O Sandy Grimes on the "backroom" operation, remarked worriedly that her good friend Rick Ames, chief of the SE Division's Western European branch, seemed to be spending a lot more money than could be justified by his salary. Ames and his second wife, a former Colombian diplomat named Rosario Casas Dupuy, had just returned from Rome and were spending money as if they were printing it themselves. They had bought a house that seemed far too dear for someone on a government salary. Rosario was redecorating lavishly, and her clothes were high-end boutique.

Grimes had been wondering about Ames, too. "He was a different human being when he came back from Rome," Grimes said later. "He had been the perfect slob, the absentminded professor. He was now a sophisticated, very proud person, erect, shoulders back, with such confidence — such arrogance. He wasn't getting any kudos from the agency. . . . He was getting them from somewhere."

Grimes took the Ames question to Jeanne Vertefeuille, who was now the chief of investigations branch of the Counterintelligence Center (CIC), a Webster innovation chartered in April 1988 to consolidate counterespionage staffs within the CIA and improve coordination with the FBI. Vertefeuille assigned Dan Payne, a CIA Office of Security investigator, to check Ames's finances. He discovered that Ames hadn't taken out a mortgage on his house; apparently Ames had plunked down $540,000 in cash for the property. That was, to put it mildly, strange. Payne queried the Treasury Department and found three wire transfers totaling about $40,000 into Ames's bank accounts. For a government employee, these were large, but there might be an innocent reason, like a bequest. But almost immediately upon assembling this information, Payne's work was interrupted when he was dispatched to "the farm," the CIA's training facility, for a routine course that took two months. After that, he was diverted to follow other tips. The Ames file would lie fallow until the spring of 1991.

Spies don't have to worry about being diverted. The down side is that, like doctors and plumbers, they often have to work nights, weekends, and holidays. Hanssen had to go to the BOB dead drop in Idylwood Park, not far from his house, on Christmas Day 1989. His Christmas present to the KGB had real sparkle: information on an NSA penetration of communications in and out of a particular Soviet facility. It might not sound sexy, but these data blew an expensive and time-consuming intercept. And there was more: a freshly minted National Intelligence Estimate titled "The Soviet System in Crisis: Prospects for the Next Two Years." There with a diskette with a file describing three new FBI RIPs inside the KGB and Soviet government, four defectors, three FBI recruitment efforts in progress, and an update on the Felix Bloch investigation. The KGB's thank-you parcel contained the obligatory Christmas greetings and $38,000 in cash, payment for the two diamonds Hanssen had returned.

In January 1990, *Pravda,* the Soviet state-controlled newspaper, reported that General Dmitri Fedorovich Polyakov had been arrested in summer of 1986 and executed for espionage on March 15, 1988. Grimes and other CIA and FBI people who had worked with the master agonized over what mistake they might have made. No one from the U.S. had heard from Polyakov since he had been suddenly recalled to Moscow from New Delhi in 1980. A CIA officer picked up some gossip in the diplomatic community that Polyakov had been diagnosed with a heart problem and had stayed in Moscow for his health. But there had long been concern that his problems were not cardiovascular in nature.

Polyakov had insisted Americans must never try to reach him in the Soviet Union, not even for a "sign of life," a signal that he was safe. In 1984, he contributed an article about hunting and cooking coot to a Russian magazine for which he had written from time to time. Years before, the CIA and Polyakov had set up an emergency communications plan: the signal would be the publication of his coot recipe. (His game-cooking tips served another purpose: by reading it a certain way, the CIA could derive a recipe for making his secret writing visible.) The article was his "sign of life" and an indication he wanted to get in touch.

Polyakov's signal set off a debate within the CIA about how to respond. The problem was, there was no guaranteed safe way to do it. The two-way burst communications device he had used in Moscow before was no longer operational. Most likely he had destroyed it because he sensed he was under suspicion and did not want to be caught with spy gear. No second signal came. Eventually top officials of the Directorate of Operations decided against attempting contact. "He had served honorably," said one CIA officer. "He was his own best judge of the situation. It was not up to us to force the issue." He had repeatedly refused offers of sanctuary in the United States. When his Delhi case officer had found Polyakov packing in 1980, he had told him, "You know, if anything happens, you are always welcome in our country." The general fixed him with steel-blue eyes. "Don't wait for me," he said. "I am never going to the United States. I am not doing this for you. I am doing this for my country. I was born a Russian, and I will die a Russian." But what will happen if he were found out, the CIA man asked. The reply came in Russian: *"Bratskaya mogila"* — an unmarked grave. The CIA took some comfort when Polyakov's son, a diplomat, was assigned to New Delhi in late 1984 or early 1985. The CIA officers figured he would not have been allowed to travel abroad if his father were suspected of espionage. But when the son was recalled to Moscow in 1986, Grimes became extremely worried. She hoped Polyakov had simply retired. She liked to think of him taking his grandchildren hunting in the Russian countryside he loved, showing them how to handle his woodworking tools. Now they all knew otherwise.

On May 7, Hanssen gave the KGB 232 pages, including additional highly sensitive information about U.S. government plans to survive a nuclear attack, the subject of documents he had sold the Soviets the previous August. For that installment, the KGB paid him $35,000, along with a diskette containing a flattering letter that inquired after his safety:

Dear Friend:

. . . We attach some information requests which we ask Your kind assistance for. We are very cautious about using Your info and materials so that none of our actions in no way causes [sic] no harm to Your security. With this on our mind we are asking that sensitive materials and information (especially hot and demanding some actions) be accompanied by some sort of Your comments or some guidance on how we may or may not use it with regard to Your security.

We wish You good luck and enclose $35,000.

Thank you.

Sincerely,

Your friends

In May of 1990, Hanssen was reassigned to the FBI Inspections Division, effective June 25, 1990. The inspection staff handled the dull but necessary work of auditing various FBI staffs for compliance with internal practices and procedures, including compliance with the bureau's security regimens. Hanssen's high security clearance, as well as his accounting experience, made him a perfect candidate for combing over the work of counterintelligence units in both domestic and overseas offices. That he was a self-righteous prig didn't hurt, either.

Hanssen informed the KGB that he was moving up in his organization, that he would be traveling for a year or so and would need ways of communicating when away from Washington for long stretches. On May 21, the KGB left a diskette with a letter in the ELLIS/Foxstone site. It read:

Dear Friend:

Congratulations on Your promotion. We wish You all the very best in Your life and career. We appreciate Your sympathy for some difficulties our people face — Your friendship and understanding are very important to us. Of course You are right, no system is perfect and we do understand this. Speaking about the systems. We don't see any problem for the system of our future communications in regard to this new circumstances of Yours. Though we can't but regret that our contacts may be not so regular as before, like You said. We believe our current commo plan — though neither perfect — covers ruther [sic] flexibly Your needs: You may have a contact with us anytime You want after staying away as long as You have to. So, do Your new job, make Your trips, take Your time. The commo plan we have will still be working. We'll keep covering the active call out signal site no matter how long it's needed. And we'll be in a ready-to-go mode to come over to the drop next in turn whenever You are ready: that is when You are back home and decide to communicate. All You'll have to do is to put Your call out signal, just as now. And You have two addresses to use to recontact us only if the signal sites for some reason don't work or can't be used. . . . But in any case be sure: You may have a contact anytime because the active call out site is always covered according to the schedule no matter how long you've been away. . . .

Thank You and good luck.

Sincerely,

Your friends

There would always be a candle in the KGB's window for "Ramon Garcia," star secret agent. Call anytime, anyplace — the toughest, meanest secret police organization on the face of the earth would be waiting with open arms, and a stack of crisp green bills.

They knew "Ramon" would be back for more.

12

THE STRIPPER AND THE SPY

Besides the spy novels he enjoyed, Bob Hanssen sporadically delved into the classics. Hanssen once told a friend that he was obsessed with the hero of Joseph Conrad's novel *Victory,* Axel Heyst. Heyst is a solitary dreamer, who believes he can avoid suffering by cutting himself off from others. Generally considered a "queer chap," Heyst starts his wanderings after the death of his father, a perpetually angry man whom Heyst feels he has betrayed.

It was more than the novel's parental issues that struck a chord with Hanssen. Heyst, confronted with a business failure on a remote island in the Malay Archipelago, rescues an English girl named Lena from an evil innkeeper and takes her to his island retreat. Lena is a young, seductive woman, "a distressed human being," determined to save Heyst from the detachment and isolation he has known all his life. She is, of course, in his debt, and can't help but acknowledge his power over her, but by saving her, he is also liberating himself. Isn't that what being God is all about?

Hanssen had fantasized about such a situation since at least his high school trips to the beach. Women hadn't exactly seen the dour, conceited Hanssen as magnetic, let alone a savior. But there is almost always a rung on the ladder low enough from which even a Bob Hanssen might look good. Hanssen found his own "distressed human being" just eleven blocks from FBI headquarters. Bob may have disdained his rowdy coworkers when they headed off to the bars and strip clubs, but, like his patriotism and devotion to the FBI, that was a bit of an act. On sultry summer afternoons, when the heat in downtown Washington was so intense that the asphalt stuck to the bottom of your shoes, Hanssen used to wander into Joanna's 1819 Club, a strip joint sandwiched between a bikini shop and a cigar store on a seedy block of M Street NW, just south of DuPont Circle. Joanna's advertised itself as "A Gentleman's Club." Inside it was anything but.

Make that butt.

Mid-afternoon was the hour Hanssen preferred, that lazy period between the lunchtime gawkers and down-to-business Happy Hour, a time when the tables were almost as empty as the eyes of the performers, and when he was less likely to be recognized by other patrons. It was also the time of day when dancers had the leisure to chat up the customers, many of whom, like Hanssen, tended to be ill-at-ease around naked women.

When Hanssen felt sure there was no one inside who looked familiar, he would take a small table with a clear view of the performers. It always took a few minutes for his eyes to adjust from the lemon-sweet sparkle of a summer day to the dark shadows of this run-down excuse for a club where the main item on the menu was strip steak. Burgundy velvet drapes, set off with tiny orange lights, flanked the mirrored walls. Red and blue spotlights bathed a low-rise platform for the dancers. In the corner, stage left, the girls kept a plastic bucket with a bottle of Windex and a roll of paper towels, which they used after each performance to wipe away sweaty body streaks made by their hips grinding into the mirrors. A deejay in the front booth chose the songs, and the performers mouthed the words.

Most of the strippers were barely a step up from the patrons, pretty low-life. They had names like Desirée, a tall chickie with thick red lipstick, a 15-watt smile, and oddly shaped, puckering breasts that could have used a shot or two more silicone. Dropping a clinging sarong, she would teeter on black, eight-inch stiletto heels, the extent of her remaining outfit, save for a tattoo of green ivy that wound down her back, an extremely uncommon piercing, and a black, elastic garter covered with glitter, holding but a few single dollar bills. It was artless, a striptease with no tease.

Loserville.

A person who saw Hanssen at Joanna's said he never really flirted with the strippers, never really spoke with them much. He just sat there, taking it all in. Nevertheless, in the summer of 1990, Bob Hanssen met Priscilla Sue Galey, a dancer at Joanna's. Priscilla recalled catching Hanssen's eye, and after she left the stage, the FBI agent gave her a ten-dollar tip with a note saying that he "had never expected to see such grace and beauty in a strip club." Priscilla was overwhelmed. It was, she said, "the most beautiful compliment I had ever heard in my entire life."

As Hanssen left the restaurant, Priscilla ran outside to thank him, and they began to talk. Soon they struck up an acquaintance. He took her to lunch and to museums. He bought her sensible shoes. Eventually, he said something that startled her, that proved he really wasn't like the other guys: he urged her to quit the skin trade. Before that had even sunk in, he encouraged Priscilla to go to St. Catherine's in suburban Virginia, his family's place of worship. Priscilla accepted, but when she got as far as the church parking lot, she saw Hanssen's family getting out of their van. Ashamed, she drove away.

Undeterred, Hanssen gave her $2,000 in cash in an envelope so she could get her teeth fixed. Then came a dazzling surprise, a sapphire and diamond necklace. Priscilla's dental work, necklace, and pocket money were all financed with the $40,000 in proceeds from Hanssen's September 4 sale of a diskette full of classified data. His exact intentions were unclear, but Priscilla insisted that they were never involved sexually. Hanssen told Priscilla that he was a happily married father of six, and even when she tried to give Hanssen an innocent hug, simply to thank him for a gift, he rebuffed her.

Maybe Priscilla filled Bob's need to show off. He knew that if he showered Bonnie with baubles, she would ask how he paid for them. "It was a great disappointment to him," said one of Hanssen's friends, "that she didn't want him to buy her expensive presents. He wanted to buy her diamonds, minks. He thought she deserved the very best, and he was frustrated that he couldn't provide them. Bonnie's idea of a great present was a new vacuum cleaner." It was a lot easier to tart up Priscilla. She didn't ask questions he didn't want to answer, and when he was done with her, he could just walk away.

Priscilla said she came to regard Hanssen as a "guardian angel," a father figure who magically appeared in her life, gave her money when she was low on cash — which was much of the time — and showed interest in her future. Priscilla's mother, Linda Harris, said that Hanssen bore an eerie resemblance to Priscilla's father. Priscilla was illegitimate and had seen this man just once in her life.

From September 1990 through February 1991 Hanssen was traveling constantly for the FBI's Inspections Division. He spent his day determining whether various field offices met the bureau's "effective and efficient" standard. He would check to see that informant payments were properly recorded, witnessed, and documented, that the offices were paying their bills on time, that they had identified the crime problems in their jurisdiction and prioritized their resources accordingly. This was an accountant's work, and he was good at it.

On the side, Hanssen kept downloading classified data onto floppy disks he smuggled out of the Hoover Building. On February 2, he resumed contact with the KGB by leaving a disk at the CHARLIE dead drop in Eakin Community Park in Vienna, not far from his home. It contained a receipt for the $40,000 payment in August. Hanssen had called it "too generous," an absurd remark, as any rookie intelligence officer would know; Soviet intelligence had seldom, if ever, received such a windfall, and at such bargain rates. Hanssen also passed along a tip he had heard from someone in the New York office — that the FBI had succeeded in recruiting a number of Russian sources in the city. That tidbit earned Hanssen another $10,000, which he collected at the CHARLIE drop site on February 18.

Almost two months later, on April 15, Hanssen went to DORIS at Canterbury Park in Springfield, Virginia, with a diskette containing another goodie: classified information about a recent FBI effort to recruit Soviet double agents. The KGB responded with $10,000 and a letter penned in the purple prose of bad novels. Hanssen's handlers had sensed something about him that the FBI had not. Flattery would get them everywhere, that it was better bait than cold, hard cash. Their letter read:

> Dear Friend:
>
> Time is flying. As a poet said:
> "What's our life,
> If full of care
> You have no time
> To stop and stare?"
> You've managed to slow down the speed of Your running life to send us a message. And we appreciate it. We hope You're O'K and Your family is fine too. We are sure You're doing great at Your job. As before, we'll keep staying alert to respond to any call from You whenever You need it.
>
> We acknowledge receiving one disk through CHARLIE. One disk of mystery and intrigue. Thank you. Not much a business letter this time. Just formalities. We consider Site-9 cancelled. And we are sure You remember: our next contact is due at ELLIS. Frankly, we are looking forward to JUNE. Every new season brings new expectations.
>
> Enclosed in our today's package please find $10,000. Thank You for Your friendship and help. We attach some information requests. We hope You'll be able to assist us on them.
>
> Take care and good luck.
> Sincerely,
> Your friends

Hanssen must have basked in the poetry and the unctuous capitalization of "You" and "Your," as if he were royalty, or a deity. Surely he chuckled at the hilarious reference to "friendship" with the KGB.

Even Hanssen would have realized that their words were hollow. But whatever the motive, this was exactly the kind of attention and adulation Hanssen craved. The KGB had his number.

That same month, Hanssen was assigned to inspect the FBI's Hong Kong office. Since the post's bread and butter was counterespionage against the People's Republic of China, he would review highly classified files having to do with the FBI's human and technical sources targeting the PRC. He was supposed to make sure agents were working on the most important Chinese espionage threats to U.S. security, that the office's physical security was adequate to prevent break-ins by Chinese intelligence agents, that the employees were being scrupulous about locking away classified documents in safes, as prescribed by government regulation, and that all expenditures, from human assets to lightbulbs, were in order.

Taking the KGB's advice about enjoying life, he impulsively invited his favorite stripper along for company. It was a high wire act for a man with a security clearance, but he took precautions. After hastily finagling a passport for Priscilla, he bought her an airline ticket on a separate flight and registered her in a separate hotel room.

Priscilla insists their relationship remained platonic, that they just met for breakfast and dinner. During the day, she roamed the city while Hanssen occupied himself at the FBI office. Conveniently, as the FBI discouraged inspectors from socializing after hours with the agents whose business affairs they were inspecting, FBI agents posted in Hong Kong never found out that the married inspector from headquarters had brought along a female companion from Washington. And not a wallflower, either, but a flashy dame who favored miniskirts and spike heels. Hanssen's in-your-face breach of ethics and security was extraordinary — and the FBI missed it completely.

On July 1, Hanssen completed his tour in Inspections and was named program manager in the Intelligence Division's Soviet section. His new job had a certain cachet. He was responsible for supervising efforts by the field offices to detect and block Soviet efforts to acquire classified scientific and technical information. KGB "requirements" lists — in common parlance, "shopping lists" — acquired through intelligence assets were about as sexy as electronics and computer manuals. Everyone in the division knew that Hanssen read that stuff for fun, so they were happy to have him take the job. It was a real promotion, the most important job he would ever hold.

Oddly, Hanssen did not hint at his elevated status to the KGB. To the contrary, in a letter left at the ELLIS/Foxstone Park drop on July 15, he sounded like an anxious lover worried about being spurned. "I returned, grabbed the first thing I could lay my hands on," he wrote. "I was in a hurry so that you would not worry, because June has passed, they held me there longer." Along with the note was a classified document describing plans by the FBI and CIA to hire and direct new double agents. This would have proved a handy guide for KGB Line KR officers charged with fending off U.S. recruitment efforts. There were also documents about the proliferation of nuclear weapons and longer-range missiles in the Third World. These reflected the high level of concern of the Bush administration and the U.S. Congress that the U.S.S.R., in economic extremis, might teach such unreliable regimes as Pakistan, Iran, Iraq, Syria, Libya, and North Korea how to make weapons of mass destruction.

The KGB stuffed the dead drop with $12,000, along with another superlative-filled note to their beloved patsy.

Dear friend:
    Acknowledging the disk and materials . . . received through "DORIS" we also acknowledge again Your superb sense of humor and Your sharp-as-a-razor mind. We highly appreciate both. Don't worry. We will not steam out incorrect conclusions from Your materials. Actually, Your information grately [sic] assisted us in seeing more clearly many issues and we are not ashamed to correct our notions if we have some. So, thank You for Your help. But if some of our requests seem a bit strange to You, please try to believe us there were sufficient reasons to put them and that what we wanted was to sort them out with Your help.
    In regard to our "memo" on Your security. Just one more remark. If our natural wish to capitalize on Your information confronts in any

way Your security interests we definitely cut down our thirst for profit and choose Your security. The same goes with any other aspect of Your case. That's why we say Your security goes first. We are sure You remember our next contact is due at "FLO." As always we attach some information requests, which are of current interest to us. We thank You and wish You the very best.

    Sincerely,

    Your Friends

The tone was cloying, but the money was green, enough for a special present for Priscilla. A really special present, one that he knew would not just thrill her but cement the notion that he was special, too. If Bob couldn't be number one on the FBI ladder, he could be a star in Priscilla's eyes. Like a man planning a proposal, and it *was* a proposal of sorts, Hanssen had to delight in every detail of it, especially the presentation. When all was ready, Hanssen took Priscilla Galey to lunch and handed her a small, tantalizing package, the kind that all women love. It was a ring — well, a key ring. With the keys to a car. And not just any car. Hanssen gave Priscilla Galey a six-year-old Mercedes-Benz 190E sedan. He had bought it for $10,500 cash. Priscilla was ecstatic. "I drove 50 miles out of the way on the way home just to drive it," she told the *Washington Post*. "I spent two weeks peeking out of my apartment windows just to make sure it was there, and it was real."

But there was more; Hanssen had thought of everything. Knowing that a working girl like Priscilla couldn't afford to keep up a luxury vehicle, he gave her an American Express card to use when she needed to pay for service. The KGB would pay the AMEX bill.

In another breach of security, Hanssen, although he registered the purchase of the car in Galey's name, used his own home address. Anyone doing a record check on Hanssen (and the FBI didn't until too late) would find that a P. S. Galey was listed as residing at the Hanssen home in August 1995. By following up the lead, the FBI might easily have found both the Mercedes and the exotic dancer who tooled around in it. Bob Hanssen would have had a few questions to answer. But he didn't. The FBI never had a clue.

Hanssen kept pressing Priscilla to leave Joanna's and get an office job. He even gave her a laptop computer so that she could learn some business skills. But there was a catch: perhaps to show off, perhaps to remind her of how much smarter he was, or perhaps just to be mean, Hanssen programmed the computer with a secret code that she would have to decipher before it would work. Not surprisingly, Priscilla, who had dropped out of high school to become an "exotic dancer," didn't have a clue as to how to work a computer, much less figure out the code; she couldn't even boot it up. Priscilla took the computer home to Columbus for the Thanksgiving holiday, hoping that her mother, Linda, who had some experience working with computers, could figure it out. But Linda was at a loss as well.

After a while it was just too frustrating and Priscilla gave up. Self-improvement was the farthest thing from her mind. Once back in Columbus, she took up with an old boyfriend, who was using and peddling crack cocaine. She became addicted to the drug herself. Before long she totaled the Mercedes, hocked the diamond and sapphire necklace and the laptop. She told interviewers that at her boyfriend's urging, she started working as a hooker. Priscilla Sue Galey would not return to Washington.

Nevertheless, Hanssen kept watching her. She found out just how closely the following April, when she used the American Express card to buy Easter outfits for her two nieces. The minute Hanssen got the bill, he flew to Columbus, charged into Linda's ramshackle house, and demanded the card's return. He might simply have canceled the card, but he wanted to make a point. Priscilla had broken the rules, and she wouldn't get a second chance. When, a couple of years later, Priscilla was arrested for dealing crack, Linda Harris telephoned Hanssen to beg for bond money. He turned her down.

13

SKYLIGHT

In April 1991, CIA counterintelligence specialists Paul Redmond and Jeanne Vertefeuille called on Ray Mislock, the FBI's new Soviet section chief, and Bob Wade, his deputy and top operations official.

The group dispensed with whatever bureaucratic issue had prompted the meeting, and the topic turned to the dark days of 1985 and 1986 when the KGB dismantled the American government's spy network in the Soviet Union. More names had been added to the list of casualties. Soviet arms control specialist Vladimir Potashov was in the dreadful prison camp called Perm 35, along with Boris Yuzhin. GRU Colonel Vladimir Vasilyev and two KGB officers, Sergei Vorontsov and Vladimir Pigusov, had been shot. At least ten Russians who had worked for the United States were known to be dead, two others were missing, and many more had been arrested, interrogated, and jailed for weeks or months. When the betrayed agents' wives and children were counted — and they should have been, for as traitors' families, they were social outcasts and reduced to poverty — the human toll was terrible.

By now, many CIA officials assumed that whoever had tipped off the KGB to the double agents was long retired. This made sense because Russians recruited after 1985 were alive and well. But Redmond suspected there was a good chance the mole was still working for the agency; and even if he or she was collecting a pension, that was no reason to close the books on an unsolved crime. "Obviously, there's a fucking spy in the place, and nobody's caught him," he fumed. "Let's get to the bottom of this."

Redmond's caustic tongue and confrontational manner had not won him many allies in the SE Division, but he had just been moved — pushed, actually — into the job of deputy chief of the CIA Counterintelligence Center. This gave him the power to do what he had been lobbying for since 1986 — take an intense, systematic look at the SE Division personnel who had access to the operations compromised in 1985 and 1986. "They had looked at the cases for commonalities," he said, "but there was a disinclination to investigate people." That was a direct result of the lingering trauma from the Angleton witch hunts, but Redmond insisted it was time to face up to the fact that cases didn't kill people, people killed people, and there was still a killer at large.

Vertefeuille, toiling in the bowels of the CIC as deputy chief of the counterespionage group, was going to retire in eighteen months, and she asked to take one last crack at the problem. Redmond promptly detached her and gave her a small office to undertake the project. It would be called SKYLIGHT.

Now Mislock sat forward. "We'd like to buy into that," he said. Though he had not had line responsibility for the COURTSHIP recruits, he felt their losses keenly, and also the losses of the CIA assets. "These people entrusted their lives to us, and they paid the ultimate price," he said later. "They were betrayed, which is different from dying in battle. That's a gut-wrenching motivation for wanting to solve the problem."

Vertefeuille was not overjoyed. She had envisioned a stand-alone shop. As far as she could tell, the FBI only got agitated when one of its own recruits was blown, and besides, it had a habit of moving in and trying to take control — which explained the bureau's nickname among smaller agencies: "the heaviest badge in town." Working with the FBI was like sharing a small tent with a bunch of big dogs.

But Redmond was nodding assent. The more eyes and hands the better, as far as he was concerned, and he liked Mislock's determination. Mislock thought that ANLACE had fizzled because the team had looked at the FBI losses as separate, small puzzles, each with its own solution. It was more likely that the compromises of both agencies' operations were parts of one big puzzle. "We had to put it all together and try to understand it in one context," he said. "We had tried to analyze our way to a solution but had not had an aggressive investigation. I thought we needed to investigate aggressively every lead, any source, anywhere." He detailed two agents to the CIA: Jim Holt, an experienced FCI street agent who had handled Martynov, and Jim Milburn, an analyst who was a walking encyclopedia on the KGB and GRU. They would occupy desks inside the CIC and have full access to all relevant CIA files.

Back at the agency, Sandy Grimes was talking about retiring. The fall of the Berlin Wall, the demise of pro-Soviet regimes in Eastern Europe, and instability in the Soviet Union itself had brought about a bumper harvest of defectors and RIPs. That was good for the agency, but it kept Grimes's back room running at such a frenetic pace that she was burning out. Office politics were nastier than usual, and her two daughters, in their teens, would rather have a stay-at-home mom than a glamorous, secret-agent mom. And she had been badly shaken by the death of Dmitri Polyakov. But when Redmond asked her to stay and work on a mole hunt, she jumped at the chance, telling him, "It's the only thing that would keep me here. I'd be *honored* to work with Jeanne. We owe those people; and we owe their families to find out what the hell happened."

By August, the task force had drawn up a long list of 198 people who had access to all the blown operations and a short list of twenty-nine people who had personal problems that might tempt them to sell out. All were Soviet/East European Division personnel. The investigators tried to be even-handed, but one name stood out:

Aldrich Hazen Ames, now working in the Counterintelligence Center analysis group. "We held a secret ballot and said, if there were a spy, who would it be, rank them one, two, or three," said Grimes. "I was the only one who put Rick Ames as number one. But you got points for being number two or three. When all the votes were tallied, Rick had the most votes."

Ames's over-the-top behavior was legendary. In 1976, he had left a briefcase full of classified documents on the New York subway. The FBI had retrieved it from a Polish immigrant who found it. He tossed clandestine communications gear around his office unsecured, couldn't account for CIA money, and often couldn't account for himself. His falling-down-drunk episodes were notorious: the 1973 CIA Christmas party in New York, when he had to be driven home by security officers; the embassy reception in Mexico City when he got into a shouting match with a Cuban intelligence officer; the time in 1987 he got plastered at Ambassador Maxwell Raab's house in Rome, passed out in a gutter, and was discovered by the police, who carted him to a hospital.

A December 1990 memo by Dan Payne of the Office of Security reported that Ames owned a half-million-dollar house free and clear, that he had "spared no expense" to renovate it, that he drove a white Jaguar, and that he had made two large cash deposits and one large currency exchange since returning from Rome. Payne urged that Ames be polygraphed immediately. But when the test was administered in April 1991, Ames passed. Moreover, in July, a CIA officer reported from Bogota that Ames's wife's family, the Casas clan, was wealthy and prominent. They might be the source of Rick's lavish spending.

"The very first day of the task force," Sandy Grimes said, "Rick walked into the office and said, 'If you need any help, any questions I can answer, just call me.' I was just about dying. I was absolutely convinced if we had a spy, it was Rick." But in some ways Ames was a little too obvious and inviting a target. FBI agents are taught not to formulate a theory and then select the facts that favor it. A theory is a heavy thing that barrels down the highway like an eighteen-wheeler. It feels solid and secure, but experienced investigators travel lighter because they know the road to the truth is full of hairpin turns and switch-backs. The agents, especially, felt it was far too soon to lock in on Ames.

Grimes, however, followed her instincts and started making a timeline of his access and movements. She queried the Directorate of Operations computer system for records bearing his name. Mean-while, the task force went down the long list of 198, interviewing, polygraphing, and assessing.

Unaware of Sandy Grimes's suspicions but annoyed by Ames's laziness and arrogance, CIA managers started bouncing him around from job to job. In September 1991, he was made chief of an SE Division KGB working group. In December, he was dumped into the CIA Counternarcotics Center. Nonetheless, he displayed an almost reck-less confidence. In January 1992, he went out and traded his old white Jaguar for a jaunty new red model.

At the FBI, the existence of the SKYLIGHT task force was kept secret. Agents working in the Soviet section knew something was up and engaged in a good deal of watercooler gossip, but there was no official confirmation. The need-to-know loop was limited to Mis-lock, Wade, and a few other very senior FBI managers — Tom Duhadway, assistant director in charge of the Intelligence Division; his deputies, Pat Watson and Skip Brandon; Robert "Bear" Bryant, the assistant director in charge of the Washington field office; and one or two others. Briefings took place away from the Hoover Build-ing, often at Watson's house. Memos having to do with SKYLIGHT were not stored in general Intelligence Division case files. They were printed out, hand-delivered to whomever they were meant for, and stored in paper files only.

Bob Hanssen wasn't in the loop. He wasn't on SKYLIGHT's short or long lists, either, because they only included CIA people. He was free to continue with the work at hand: selling whatever strategic and technological advances he could get a hold of. On August 19, 1991, Hanssen went to the drop site called FLO, under the footbridge at Lewinsville Park in McLean, and deposited, among other items, TOP SECRET/SCI information revealing exactly how NSA was reading certain communications. Hanssen could not have chosen a worse time to pass information that could blind U.S. intelligence to an important Soviet government communications channel. The day before, KGB chief Vladimir Kryuchkov and other hard-liners opposed to Soviet leader Mikhail Gorbachev's reform initiatives — *glasnost,* openness toward the West, and *perestroika,* economic restructuring — had mounted a putsch, ordering Gorbachev detained at his holiday retreat in Crimea. At six A.M. August 19, Moscow time — ten P.M. August 18, Washington time — the coup leaders announced via Tass, the Soviet news agency, that Gorbachev was ill and had surrendered his powers to one of their number, Vice President Gennadi Yanayev. A short time later, Tass reported the government was now controlled by the State Committee for the State of Emergency, as the plotters called themselves. Tanks rolled in the streets, demonstrations and political parties were banned, curfews were decreed, and independent newspapers were shut down. The stakes could not be exaggerated. For seventy-six hours, the junta operatives who put Gorbachev under house arrest had the "football" — the briefcase with launch-authorization codes. "We didn't know who was in charge," said a former U.S. State Department expert on Soviet policy. "It was clear Gorbachev was not. The people who were in charge for those days were retrograde and, we had reason to believe, drunks." If the U.S.S.R. fell apart, what would happen to the 27,000 nuclear warheads sprinkled across several republics? Societies that had been closed for decades were hard enough for the West to understand. Bob Hanssen was shattering the few windows the American intelligence community had into this volatile part of the world.

At the moment when Hanssen was making the drop at Lewinsville Park, pro-democracy leaders were calling on Muscovites to take to the streets. By noon Tuesday, the twentieth, some 150,000 Russians gathered in front of the headquarters of the Russian Republic (known as the "White House") to hear Boris Yeltsin declare, "We will hold out as long as we have to." Major General Alexander Lebed, commanding the tanks surrounding the White House, broke with the junta Tuesday afternoon and ordered the tank turrets turned around. Other military leaders also refused to storm the White House, and by the morning of Wednesday, August 21, the coup was over, and the plotters had been arrested. Kryuchkov, who had so lovingly pored over Hanssen's reports, was charged with treason, interrogated, and imprisoned. As the coup fell apart, a jubilant throng surrounded KGB headquarters, and a crane was brought in to drag the statue of KGB founder Felix Dzherzhinsky from its pedestal.

In his correspondence with the KGB, Hanssen was way behind the news curve. Not only did he not know about the coup, but — blowhard that he was — Hanssen enclosed a letter encouraging the KGB to study the inner workings of Chicago mayor Richard J. Daley's political machine for tips on how to maintain power. His handlers probably laughed mightily when they saw that their spy — who had proven so easy to manipulate — revealed such utter ignorance of how the Soviet Union, let alone America, actually worked. (In their next communication they would mock his arrogance, and still Hanssen would take their over-the-top praise as legitimate.) The KGB officer who collected his offering left him a polite letter of thanks and $20,000.

A few weeks later, during Secretary of State James A. Baker's visit to Moscow, Vadim V. Bakatin, a reformer named by Gorbachev to carve the KGB to bits, boldly proposed that the two nations limit spying on each other and asked for CIA help to draft rules that would help reinvent the Russian intelligence service to fulfill its mission in a democratic society. But Bakatin's promises, if earnest, were ignored by the spies in the States. During the coup and in the months that followed, FBI agents in the Washington field office counterespionage squads reported that dead drops were being serviced, coded messages dispatched and received, clandestine meetings held, all as usual. The only difference was, the KGB got a new name, Sluzhba Vneshney Razvedki Rossii, or SVR, but Russian spies carried on business as usual in Washington, New York, and San Francisco. Tom Duhadway, regarded as the finest strategic thinker the Intelligence Division had ever produced, went to work on a plan to keep the pressure on the

Russians but at the same time bring new resources to bear on the post–Cold War threats of terrorism, economic espionage, and proliferation of weapons of mass destruction. Sadly, he would not live to carry it out. Duhadway died of a heart attack on September 21, 1991, at the age of forty-nine.

Even as the map of the world changed, Hanssen and his Russian contacts maintained an oddly normal routine. On October 7, the FBI agent left a package at the GRACE drop, a footbridge in Rock Creek Park in Washington, with information about FBI recruitments and a classified report, *The US Double-Agent Program Management Review and Policy Recommendations.* This report was one of Hanssen's most important contributions to the SVR. It offered insights into exactly how the FBI's counterintelligence program functioned. As had been the case in the past, though it did not contain true names of FBI recruits, the descriptions were sufficiently precise that any Russian officer who read it could discern the identities of the double agents. As well, the document described what the FBI planned to do in the future — who would be approached, when, and why.

At the same time, Hanssen made a move that was as reckless as his decision to take a stripper to Hong Kong. Only several days before, his best friend Jack Hoschouer had confessed to Hanssen that he had been passed over for promotion, a signal that he would be forced into retirement without a chance of wearing a general's stars. Hanssen seized on the opportunity. He urged the Russians to recruit Hoschouer, who was now a lieutenant colonel in the U.S. Army and stationed in Germany. Having committed what would be argued were capital crimes, he now attempted to take his best friend down with him.

The Russians filled the drop with $12,000 cash and a letter:

Dear friend:
  Thanks for the package of 02.13. [The] materials are very promising, we intend to work on the scenario so wisely suggested by You. And the magical history tour to Chicago was mysteriously well timed. Have You ever thought of foretelling the things? After Your retirement for instance in some sort of Your own "Cristall [sic] Ball and Intelli-

gence Agency" (CBIA)? There are always so many people in this world eager to get a glimpse of the future.

But now back to where we belong. There have been many important developments in our country lately. So many that we'd like to reassure You once again. Like we said: we've done all in order that none of those events ever affects Your security and our ability to maintain the operation with You. And of course there can be no doubt of our commitment to Your friendship and cooperation which are too important to us to loose [sic]. . . .

Sincerely,

Your friends.

Among the items Moscow Center wanted Hanssen to provide was an issue of an internal intelligence community publication about U.S. satellite reconnaissance systems, which would reveal the level of detail captured on film by the classified high-resolution cameras deployed aboard the spy satellites. The Russian military could use this knowledge to improve its camouflage of weapons plants, airfields, missile silos, command centers, and other military installations. It could also trade the information to pariah nations like Iran, Iraq, Libya, and North Korea. The SVR didn't say that, of course. Instead, the words bordered on the ridiculous. "It's fun to read about the life in the Universe to understand better what's going on our own planet."

In a matter of days, a delegation of pro-democracy Russians involved in reforms of the Russian intelligence services arrived in Washington. The delegation, led by Sergei Stepashin, chairman of the Russian republic's parliamentary committee on state security, toured the FBI and CIA, the U.S. Customs Service, the Drug Enforcement Administration, and Capitol Hill. The irony of the two spy communities meeting in the name of espionage reform was delicious, and there was much glad-handing and toasting. In a scene that would have horrified J. Edgar Hoover, the Russians had their pictures taken with FBI director William Sessions and his top aides. One of the Russians was Alexander Sterligov, a reserve KGB general who had helped Gorbachev escape the coup leaders.

Stepashin told George Lardner of the *Washington Post* that his group had raised the idea of intelligence-sharing with the CIA; he spoke of creating "informational centers" in Moscow and Washington where data on terrorism, drugs, politics, and economics could be exchanged. Acting CIA director Richard J. Kerr told the *Wall Street Journal* the agency would enter serious discussions with the Russians. "We have a new KGB," Kerr said, "and I think we have to be willing to look at this KGB with new eyes, just as we're looking at the Soviet Union."

The talk of cooperation between the two spy agencies didn't faze Hanssen at first. On December 12, 1991, he coolly dispatched a letter with details of an FBI electronic surveillance plan targeting the Russian Embassy.

Four days later, on December 16, Hanssen filled a dead drop named BOB in Idylwood Park, between Vienna and Falls Church, with an abundance of top-grade classified data: a TOP SECRET/SCI report about a highly sensitive NSA intercept; a SECRET research paper from the Counterintelligence Center titled *The KGB's First Chief Directorate: Structure, Functions, and Methods,* and dated November 1990; and a SECRET budget document that detailed the FBI counterintelligence program's resources, personnel, deployments, and goals.

A letter that Hanssen placed on a diskette boasted of a coming promotion. "A new mission for my new group has not been fully defined," he wrote. "I hope to adjust to that. As General Patton said, 'Let's get this over with so we can go kick the shit out of the purple-pissing Japanese.'"

The new job Hanssen alluded to was chief of the National Security Threat List (NSTL) Unit. The unit, conceived by Duhadway and established by his successor, Wayne Gilbert, was to focus the FBI's response to post–Cold War national security dangers such as terrorism and proliferation and to redirect FBI counterintelligence resources to meet them. Hanssen was to pay special attention to economic espionage by both friendly and hostile nations. This was an important advance in terms of prestige and salary. As a GS-15, he would make about $80,000. Once again, it was not the sort of job a field agent would want, but a perfect place for a spy. "People say he was of no value to the Russians there, but he could get great information," said Skip Brandon. For example, Hanssen could have offered the Russians information about how nations like France and Japan were being targeted by the United States for economic counterespionage. The Kremlin could use these data to undermine American alliances.

Hanssen wrote that while he would not have direct responsibility for Russian affairs, he intended to continue supplying information covertly. He was mulling setting up an office in downtown Washington to communicate securely with the Russians by means of a special computer encryption system. He added that he was confident the office would not be bugged.

And suddenly, he stopped. Over the night of December 16, the Russians replaced Hanssen's package with a packet containing $12,000 cash. This transaction appears to have been the last between "B" and Russian intelligence for eight years.

Why did Hanssen pull back? Some of his colleagues believe he saw the Russian delegation visiting FBI headquarters and realized that the two agencies might actually establish some sort of liaison. Five days after the $12,000 pickup, Boris Yeltsin and the presidents of Ukraine and Belorussia had declared the U.S.S.R. dead. Cooperation was not out of the question. Even if the official exchanges were limited to "safe" topics like terrorism and drugs, eventually personnel from the two agencies would bond. That almost always happened among law enforcement and intelligence people of any nationality. Beer and vodka would flow and so would war stories and gossip. Sooner or later, some Russian might start talking about the great "Ramon Garcia." One conversation would lead to another and before long, one of Hanssen's colleagues would knock on his door and read him his rights.

Or maybe Hanssen realized that the upheaval in Moscow would drive dozens more KGB and SVR officers into the arms of the CIA, FBI, and other Western intelligence services. He must have heard

Mislock and the others repeating the maxim: Spies catch spies. Would some ex-KGB defector do to him what he had done to Polyakov, Martynov, Motorin, Yuzhin, and others he had betrayed? In a heartbeat.

Even so, spying was a hard habit to break. He had been thinking about quitting for some time, but he didn't think he could do it on his own. "He needed the assistance of a priest," said someone in whom Hanssen confided after his arrest. Hanssen did not want to go to a priest in the Washington area, nor to anyone associated with Opus Dei. Instead, as Hanssen told the story, while on an inspection tour in Indianapolis, he walked into a Catholic church near the FBI office, slipped into the confessional, and told the priest he had been a spy but that he wanted to quit. Whatever else was said remains between Hanssen and the priest.

Though Hanssen would later cite the priest's words as the push he needed, they clearly didn't stop him. Nothing really did. He would be patient, but he wasn't going to wait forever. It was time to sleep, but just for a while.

14

PLAYACTOR

By the end of 1991, SKYLIGHT was starting to take on a faint glow. The team was still plowing through the long list of 198 CIA employees, reviewing records, and running down leads that in a couple of cases proved to be major wastes of time. But the two cultures seemed to have meshed, finally. "Jim squared," as the two FBI men were called, were getting along well with Vertefeuille and Grimes and had no complaints about being shut out of meetings or denied documents. Relations between the FBI and CIA were so warm that Milt Bearden, chief of the agency's Central European Group (the renamed Soviet/East European Division) had asked to borrow Bob Wade to be his counterintelligence advisor.

Ray Mislock still wasn't satisfied. He had no doubt that if there was a mole in the CIA clandestine service, sooner or later SKYLIGHT would find him, or at least put together enough information to open a criminal investigation. But Mislock trusted his instincts, and they were telling him that SKYLIGHT's scope was too narrow. What if the mole wasn't at the CIA? What if there was one double agent at CIA but another, or several, at the Defense Department, the State Department, or even, God forbid, the FBI? As tragic as the human losses were, there might have been other compromises that had not been recognized because they involved classified analyses, scientific and engineering data, or the Pentagon's technical collection programs.

Mislock took his misgivings to his boss, Deputy Assistant Director Pat Watson, who had spent twenty-four years in Soviet counterespionage. Watson agreed. So they started up yet another secret inquiry, aimed at understanding the universe of leads from all sources in the broadest possible context.

The new project was called PLAYACTOR. Its mission was to find out everything there was to know about Soviet recruitment in the United States. "From that we would begin to reduce the background noise," said Mislock, "and assemble a team of our most qualified agents who would become the experts on the problems. The most promising leads would become investigations and aggressively pursued." Whenever enough evidence was developed against a particular suspect, a criminal case would be spun off. But there would always be a core group of investigators to stay behind to look at what had not been resolved.

To run PLAYACTOR, Mislock tapped Tim Caruso, chief of analysis for the Russian counterintelligence section, which had replaced the Soviet section at FBI headquarters. Caruso, a lean, angular man with chilly blue eyes, had a bit of Inspector Javert in him. His colleagues insisted he was not obsessive and cruel, like the policeman in Victor Hugo's *Les Misérables,* but they agreed they never, ever wanted to be in his crosshairs. Mislock liked the way he approached every case "with a sense of urgency."

Investigations belonged in the field, not headquarters, so Watson and Mislock asked Bear Bryant, the assistant director in charge of the Washington field office (WFO), to give PLAYACTOR space in their building, a dilapidated structure on Buzzard's Point, one of the bleakest, most dangerous neighborhoods in Washington. Bryant was relieved. A big, blunt ex–football player from Missouri, he had spent his whole career on criminal investigations and knew next to nothing about FCI work. He had not rested easy in months, ever since Tom

Duhadway had taken him to dinner, glanced about to make sure no one was within earshot, and whispered, "There's a major penetration in American intelligence." Once a suspect was identified, the criminal investigation would be the responsibility of WFO, meaning Bryant. Surveillances would be mounted, records scoured, wiretaps and searches conducted, all without the smallest breach of secrecy. It was going to be the most important espionage case in decades. Another Howard or Bloch fiasco would be ruinous for the U.S. intelligence community — and a serial killer would escape. Bryant tried not to think about what would happen if he blew it.

When Caruso arrived to set up the team in January of 1992, Bryant gave him the big corner suite that had just been vacated by the task force that had solved the Pan Am 103 bombing. Caruso lined the walls with great sheets of butcher paper and began to create an elaborate timeline going back decades. It showed every case, every lead, every piece of technical and human evidence the bureau had ever uncovered. The spies lost in 1985 and 1986 were there, along with a long list of lesser operations that had gone bad. Names of some people who knew about each secret operation and who exhibited "risk factors" were added as the information developed.

Caruso called the work in progress a matrix. It was a graphic representation of the investigator's mantra: *motive, opportunity, and means*. Where opportunity (access to a secret) intersected with a motive (financial need or a desire for revenge) or a means (access to a Soviet buyer) were the points at which to concentrate the search. The matrix was classified Top Secret and rolled up and locked away every night. It was dauntingly complex, but when Caruso briefed officials at headquarters, he transformed the lines, whorls, and arrows into a rich narrative with intricate plots and subplots.

The major subplot, the SKYLIGHT investigation, made its first breakthrough in June 1992. Dan Payne of the CIA Office of Security had joined the task force and had resumed his financial investigation of Aldrich Ames. He obtained copies of Ames's credit card and bank records, which showed the intelligence officer had been spending an astonishing $20,000 to $30,000 a month. They also disclosed that

Ames had made a number of trips abroad without reporting them to his superiors, as required under CIA rules. Sandy Grimes plugged the dates of Ames's large cash deposits into the timeline she had been constructing. Several occurred the day after Ames had reported meeting a Soviet diplomat named Sergei Chuvakin, whom he had claimed to be attempting to recruit. The team followed some wire transfers back to an account at Credit Suisse in Zurich. When these records were obtained in the fall of 1992, the investigators learned that Ames had banked well more than $1.3 million over and above his salary. They also learned that he had received an inheritance and that the Colombian relatives of his wife, Rosario, while socially prominent, were not wealthy. The paper trail Ames had left was not hard proof that he was a spy, but what other explanation could there be? Given his obvious intelligence, it was hard to understand how he could have been so sloppy.

"I don't think he thought he was going to be caught, not by Jeanne and Sandy," Grimes said. "He thought we were dumb broads, and he was far smarter than we were." But there was a long way to go before a case against Ames would stand up in court.

While the mole hunters toiled away in darkest secrecy, back at the Hoover Building, tensions were running high. Adjusting to the post–Cold War world in practical terms was a stressful task. The FBI, like the CIA, was deluged with would-be defectors who had worked for the KGB, GRU, and Bloc intelligence services. The files of the Stasi, the East German state intelligence service, had been flung open, and these had to be studied for evidence of double agents, going back decades. SVR director Yevgeni Primakov was continuing to meet with various U.S. officials about possible cooperation. The talks were tentative, but who knew where they might lead?

Hanssen, in charge of the Threat List unit, annoyed his superiors more than usual with his poking about. "Bob was a hall-walker, always sticking his nose into a meeting," Skip Brandon said. "I once looked up and saw him and said, 'You're not required in this meeting. Please leave.' He always seemed to be lurking around." Brandon and others wrote off his inquisitive ways to his usual obtuseness; being an insensitive clod was good cover. Sometime in 1992 or early 1993, Hanssen started complaining about the division's new computer network. It was a local area network, or LAN in computer-talk, not linked to the network that served the rest of the bureau. The systems people said it was totally secure, but Hanssen carped that they didn't know what they were talking about. His colleagues and superiors ignored him. They knew he was more skilled at the computer than the rest of them combined, but he never had much good to say about anything, so his complaints did not rise above the level of white noise. "Bob had troubles presenting his ideas in a way that was effective," said Brandon. "There is a way to do it without alienating people in your area of responsibility. He was a thoughtful, bright guy, but he wrote offensive memos. People had the feeling of, sigh, here's another memo from Hanssen."

One day, Hanssen marched into Mislock's office and thrust a piece of paper in his face. "Here, take a look at this," he snapped. It was a copy of a classified memo on a very sensitive subject that Mislock had written just moments earlier. Somehow, Hanssen had sat down at his own keyboard, found a path into Mislock's hard drive, copied the memo, stored it on his own computer, and printed it out.

Mislock's face colored a deep red. The most sensitive files about the PLAYACTOR and SKYLIGHT investigations were not stored in the network or on Mislock's hard drive, but his computer held references to the existence of these groups, information that would have been very useful to someone seeking to know what the bureau was up to and evade its reach.

Mislock was, as he put it later, "ripshit." He snatched the paper away from Hanssen and ran down the hall to Watson's office, where he started venting about the lousy systems and the computer technicians who had sworn that this kind of thing could never happen. Watson and Mislock went in search of Brandon, who was the liaison to computer support. Hanssen trailed behind. As Brandon asked questions, trying to understand what had happened, Hanssen went back to his cubicle, made a few keystrokes, and returned with a memo that Brandon had just written.

The decibel level soared. The computer people were summoned and upbraided. Hanssen took them into his office and demonstrated how he had hacked into the other computers. They promised to correct the problem. "I'd heard it before," said Mislock, and had all of his section's computers unplugged from the network. Brandon thanked Hanssen for his help but admonished him not to do it again.

It was not until Hanssen was arrested, nearly a decade later, that Mislock, Watson, and Brandon all realized that they had been had. Hanssen, they figured, had probably gotten a whiff of the mole hunts and started looking around in the network files to see if he was a target. He knew that a smart computer specialist would be able to follow his trail. And being a smart computer specialist, he invented a smart cover: his concern about the security of the system. "What he did was disarming," said Mislock.

On one occasion, Hanssen lost control of his emotions altogether. In February 1993, when an intelligence analyst named Kimberly Lichtenberg walked out of a boring meeting, Hanssen went after her, grabbed her arm, and yanked her violently back toward the room. She twisted around and fell. Lichtenberg filed a lawsuit against Hanssen, which was later dropped because she failed to make a court appearance, but the incident was telling. Hanssen was investigated by the personnel office and suspended for five days without pay. No security clearance review was done, though it now seems clear that Hanssen was under enormous pressure, and the FBI should have asked why.

In the glare of hindsight, Brandon and others who knew of the incident wish that Hanssen had been disciplined much more severely and referred for a psychological examination. But FBI managers lagged way behind corporate America in recognizing workplace violence as a pervasive and serious phenomenon. Moreover, sending an FBI agent for evaluation might cause him to be relieved of his bureau-issue sidearm, an act that was, at least symbolically, emasculating. Consigning someone to the "rubber gun squad" was a step FBI executives were reluctant to take. "Once you were an agent, you were supposed to be trusted," said Clint Van Zandt, an FBI behaviorist, profiler, and hostage negotiator, who was based at the FBI Academy in those days. "But the bureau was a microcosm of society, and many times inappropriate behavior would be overlooked, at least the first time. The thought was, the agent was under stress and would be able to deal with that stress and move on with his life. It was, okay, he had a bad day, and we'll cut him slack." Van Zandt, who retired from the bureau in 1995, said that Hanssen's outburst "today would be seen as requiring either dismissal or mandatory psychological evaluation and anger management counseling. It was a red flag that could have covered the whole J. Edgar Hoover Building, but for whatever reason, they ignored it." If he had been asked, Van Zandt said he would have told headquarters, "People don't just snap. This is long-term behavior, something that's been festering for quite a while."

But he was not asked, and there is no evidence that anybody else in the behavioral science unit was consulted, nor that an evaluation from one of the bureau's retained psychiatrists was sought.

Maybe what Hanssen really needed was a self-help center, a new version of an old-model Spies Anonymous. Going cold turkey was a lot tougher than he had anticipated. But with both the FBI and CIA avidly scouting out ex-KGB men around the world, he didn't dare send Ramon Garcia back into action.

One day in 1993, Hanssen came up with a plan. Let Ramon sleep. Invent a new character. In fact, start the game all over again — this time around, with the GRU. The Russian military intelligence agency was small but relatively stable, untroubled by the demoralizing "reforms" visited on Moscow Center after the KGB leadership's involvement in the disastrous putsch of August 1991.

Hanssen fixed upon a particular GRU officer, waited outside the Russian Embassy on Wisconsin Avenue for him, followed him to a parking lot, and approached him, offering to peddle government secrets. The startled GRU officer took one look at the glum American in his white shirt, conservative tie, and cheap dark blue suit and fled. Apparently, he sized Hanssen up as a "dangle," sent by the FBI to entrap him in a criminal conspiracy that would end with a triumphant FBI press conference — and for the hapless Russian, a one-way ticket back to his cold, gray, bankrupt homeland. No thank you.

Hanssen's performance was careless and sloppy. In the past, he had acted with great calculation. His failure to think things out smacked of desperation. He was lucky that he didn't get caught. At that very moment, the streets of Washington were filled with FBI agents hunting for a Russian mole.

In March 1993, the SKYLIGHT team had completed a report naming Aldrich Ames as a prime suspect in the 1985–86 losses, and the FBI Washington field office had opened a full field criminal investigation of the CIA officer. John Lewis, an ex-Marine who was special agent in charge of the field office's Intelligence Division, was supervising the case. Les Wiser was the case agent, with day-to-day responsibility for the investigation. Wiser called the Ames case NIGHTMOVER.

The FBI obtained warrants under the Foreign Intelligence Surveillance Act, tapped Ames's phones, monitored his mail, watched his car's movements with a covert tracking device, searched his house and cars, and downloaded the data on his computer. Louis Freeh, sworn in as FBI director on September 1, 1993, was briefed on the investigation the very next day. Appalled, he demanded briefings at least weekly, and, after a while, every day or two. He was impressed by Bryant and promoted him from WFO to be the assistant director of the headquarters Intelligence Division (which Freeh renamed the National Security Division), its mission expanded to include more authority for terrorism investigations.

On September 15, 1993, Wiser's team, in a search of Ames's trash, turned up a yellow sticky note referring to a signal site. That suggested he was actively filling dead drops. An October 12 wiretap caught him telling his wife he planned to "put a signal down . . . confirming that I am coming." Wiser and his team wanted to catch Rick red-handed at a drop site, but he was always a step ahead of them. Finally, they arrested Rick and Rosario Ames on February 21, 1994, just as he was about to leave for an official conference in Moscow concerning international cooperation on drug control.

Aldrich Ames pleaded guilty to espionage on April 28, 1994. His lawyer, Plato Cacheris, who had long experience in espionage cases, worked out a plea bargain in which Ames would get life in prison and a reduced sentence for his wife in exchange for telling everything he had done.

Ames began by confessing that he had begun spying on April 16, 1985, when he walked into the Soviet Embassy on Sixteenth Street in Washington and left a letter for Stanislav Andreevich Androsov, the KGB resident, meaning, the KGB's senior official in the United States. "Seeking money to pay debts," he said, "I conceived a kind of confidence game to play on the KGB. In exchange for $50,000, I provided the KGB with the identities of several Soviet citizens who appeared to be cooperating with the CIA inside the Soviet Union. I suspected that their cooperation was not genuine, that their true loyalty was to the KGB and, therefore, I could cause them no harm." In other words, Ames arrogantly asserted that his first act was harmless because it was based on his "suspicion" that the people he betrayed were KGB double agents.

"Then, a few months later, I did something which is still not entirely explicable to me," Ames went on. "Without preconditions or any demand for payment, I volunteered to the KGB information identifying virtually all Soviet agents of the CIA and other American and foreign services known to me." On June 13, 1985, he went to Chadwick's restaurant in Georgetown and handed a bag, packed with as much as seven pounds of documents, to Sergei Chuvakin, the Soviet diplomat who served as the cutout between Ames and KGB Line KR chief Viktor Cherkashin, a wily pro well known to the FBI because he oversaw all penetrations achieved by the residency. In the bag were papers identifying Polyakov, Martynov, Motorin, Yuzhin, and nearly all the other agents lost in 1985 and 1986. He said he had simply stuffed the documents into bags and walked out of the CIA. No one checked.

Bob Hanssen, the pious FBI agent and devoted family man, knew that Ames's betrayal of the Russian spies was the luckiest thing that ever happened to him. Hanssen had betrayed at least four of the same men fingered by Ames. Polyakov, Martynov, and Motorin, who had been executed, and Yuzhin, who had only escaped death in the Gulag because President Boris Yeltsin declared an amnesty for political prisoners in 1992. Hanssen was a second source, a very reliable source indeed, who confirmed the erratic Ames's tales and may have sealed the fate of Martynov and Motorin. As for Polyakov, Hanssen was the original source, but it was possible that the GRU buried his tip because its leaders hated and feared the KGB more than they hated the United States. Ames was condemned as a cold-blooded murderer, who had reaped $2.5 million for sending ten men to their deaths, ruining many other lives, and undermining U.S. security.

The Ames scandal caused CIA officials to be pilloried in the press and on Capitol Hill for harboring the shiftless lush and failing to spot the warning signs of a traitor. CIA inspector general Frederick Hitz recommended that twenty-three current and former CIA officials be held accountable for "gross negligence." CIA director James Woolsey issued letters of reprimand to just eleven of these, a decision that touched off a firestorm, marring his last months in office. The Senate Select Committee on Intelligence praised the dedication of the SKY-LIGHT task force but criticized it for slowness and timidity.

In public and private, FBI director Freeh and his top advisors seldom missed a chance to burnish the bureau's image at the expense of the CIA. A 1997 report by Justice inspector general Michael Bromwich asserted that the fault was not entirely the CIA's. "Throughout nearly the entire nine-year period of Ames' espionage, FBI management devoted inadequate attention to determining the cause of the sudden, unprecedented, and catastrophic losses suffered by both the FBI and the CIA in their Soviet intelligence programs," Bronwich found. "Indeed, FBI's senior management was almost entirely unaware of the scope and significance of the mid-1980s losses and of the FBI's limited efforts to determine their cause. FBI senior management's lack of knowledge concerning the intelligence losses contributed to the FBI's failure to devote priority attention to this matter, particularly after 1987. Moreover, the FBI never showed any sustained interest, prior to mid-1991, in investigating the enormous intelligence losses suffered by the CIA. Even when a joint effort was initiated in mid-1991, that effort suffered from inadequate management attention as well as insufficient resources."

Despite these stinging words, the Bromwich report didn't help the CIA, and it certainly didn't hurt Freeh, because the failure Bromwich cited had occurred during the tenures of William Webster and William Sessions. As was his habit during his eight-year term, Freeh shifted the blame to his predecessors and to faceless careerists. He positioned himself as a reformer poised to transform the short-sighted, reactionary bureau culture, pledging to forge strong bonds with the CIA, to employ more and better analysts, and to pick coun-terintelligence managers adept at the subtle job of spotting moles before they inflicted mortal damage on the system. This was just what congressional leaders wanted to hear. Freeh was rewarded with vastly more political clout than the CIA. The FBI would dominate counter-intelligence policy into the next decade.

But inside the FBI, the old hands in what was now the National Security Division weren't popping any champagne corks. Ames's confessions explained a lot, but there were some important things he could not have given away. He had not been in a position to know about the Felix Bloch case, nor to alert Bloch or Reino Gikman, his KGB contact. His voice was not the voice of "Ferdinand Paul," the mystery caller who had warned Bloch away from the fateful meeting at which the FBI and CIA had hoped to nail down proof that Bloch was a KGB RIP. There were other CIA losses that didn't fit Ames's access and movements. Also, sometime in the spring of 1993, Caruso told his superiors that the PLAYACTOR matrix showed a cluster of compromised operations in the New York field office between 1987 and 1990.

"When we got through digesting all of Ames, we knew there was more work to be done," said Mislock. "The potential was that there was someone even more valuable to the KGB than Ames."

The hunt was still on.

15

BLINDSWITCH

At the same time Louis Freeh was being briefed on the Ames case, Tim Caruso walked the new director through the PLAYACTOR matrix, which had grown to four or five feet high and seven to ten feet long. Caruso dwelled on the odd group of apparent compromises in New York between 1987 and 1990. Nearly all the losses were incurred by Squad I-9, which specialized in illegals. A few more losses had been suffered by the squad in the neighboring space. None of the bureau's sources had been executed, as far as Caruso could tell, but a number of very sensitive operations had been rolled up. Efforts to get close to certain Soviets had failed because the surveillance targets didn't show up where they were expected. These losses appeared to have nothing to do with Aldrich Ames.

Caruso did not know whether the penetration was technical or human but he was convinced that one existed. Jim Fox, the assistant director in charge of NYO, was not so sure. Fox, who had made his mark running counterintelligence operations in the San Francisco

FBI office in the late seventies and early eighties, argued that the blown operations could just as easily be attributed to accidents or bad luck. There was some information from Russian sources supporting the penetration theory, but Fox countered that this might be disinformation from the SVR. Still, Caruso maintained that the penetration was real. There was a furious debate. Fox and Caruso both had impeccable reputations, but in the end, Freeh was inclined to go with Caruso's darker scenario.

The New York office was like a second home to Freeh, the place where he had met his closest friends. Also, Freeh wanted no blotch on his record. He was brainy, disciplined, coldly ambitious, and a cunning bureaucratic politician. One quality he did not possess was patience — he wanted the case solved yesterday, and when it wasn't, he micromanaged. In December 1993, Freeh decided Fox's team wasn't making sufficient progress, so he called in Tom Pickard, who was the New York office's special agent in charge for administrative affairs. Freeh and Pickard had been street agents together in the late seventies. The aloof Freeh and the genial Pickard were nothing alike, but Freeh thought Pickard was a good leader. "Have you ever worked counterintelligence?" Freeh asked. "No, not a day in my life," Pickard replied. "Good, then you're not a suspect," Freeh said. "Congratulations, you're in charge." With that, Pickard took command of the National Security Division in the New York office. His mission from Freeh was to do whatever was necessary within the law to find the problem and fix it. Yesterday.

Pickard found the case, which went by a computer-generated name BLINDSWITCH, to be the most formidable he had ever encountered. Coming up with a short list of suspects was a huge job: the mole could be any agent who had served on the two squads during the four years in question, and since the FBI transfers agents frequently, the number of NYO alumni amounted to about two hundred agents. The long list also included non-agent personnel such as surveillance specialists, analysts, systems people, the "wires and pliers" guys in special operations, secretaries, and clerks. Then there were those who might have overheard something — carpool mates, drinking buddies, workout partners. And there were headquarters people who had reviewed the reports of the NYO squads.

By late spring of 1994, when the investigators had nothing but weak circumstantial evidence, Pickard lobbied headquarters to assign Tim Caruso to New York as his assistant special agent in charge. Caruso was full of fresh ideas. One of PLAYACTOR's goals had been to identify and target for recruitment all current and former Russian intelligence professionals who knew the answers to the penetrations of the U.S. government. So far, neither the FBI nor CIA had landed any of them. There was no shortage of ex-KGB officers and other washed-up Soviets down on their luck and eager to snitch on the Yeltsin government and the SVR leadership. But not just any old Soviet spy would do. The KGB veterans who knew about American moles composed a tiny circle, and so far, none of them was willing to change sides. Maybe Caruso could help.

But that didn't mean there weren't other Russians who knew something. In early 1995, the BLINDSWITCH squad came up with the name of Rollan G. Dzheikiya, who had been a senior counselor at the Soviet Mission to the UN. He had not been KGB, but his job, handling accreditation for Soviet diplomats and liaison with U.S. authorities, caused him to know the spies in the residency. Downsized after the dissolution of the Soviet Union, Dzheikiya wanted to stay in New York. He felt badly treated by the Russian government; his wife had left him and returned to Moscow; and he needed a green card and a job. When the FBI agents approached him with a proposition, he smiled. "What took you so long?" he said.

In Dzheikiya the agents had hit the jackpot. The Russian said that in July 1987, he received a letter from an FBI agent, saying he had seen Dzheikiya at JFK International Airport greeting a known KGB officer. Assuming Dzheikiya was also KGB, the FBI agent offered to sell information and included a sheet from an FBI administrative list to prove his bona fides. If the KGB was interested, the agent wrote, they could meet at the New York Public Library. Dzheikiya gave the letter to a security officer, who gave it to Aleksandr Vasilyevich Karpov, a KGB Line KR officer.

Two weeks later, Dzheikiya met the American at the library and introduced him to Karpov. The KGB told him to forget about what he saw, and he never learned the American's name. When the FBI agents showed him a photo lineup, he picked out the picture of Earl Edwin Pitts, who had served on Squad I-9 from early 1987 through part of 1989 and was then teaching criminal psychology at the FBI Academy's behavioral sciences unit in Quantico.

Earl Pitts did not fit the FBI's profile of a spy. He was a former Army captain with master's and law degrees. His colleagues regarded him as a quiet, responsible, straight-shooter who preferred pushing paper over shoe-leather work. "Corn-fed," one friend called him. He was neither a drinker nor a drug abuser. He was a loner, but he seemed mentally stable. He had never been in trouble. Working in the New York office was rough on everybody, but Pitts and his wife, Mary, an FBI administrative employee until 1992, had no children, lived modestly, and did not seem desperate for cash.

Because a big part of the investigation had to be done in Virginia, where Pitts lived, Mislock, who had become special agent in charge of the National Security Division at the Washington field office, and Les Wiser, whose counterespionage squad had run the Ames case, conducted an extensive financial analysis of Pitts's assets and bank accounts. They did not turn up any hidden caches of money. His travel and telephone records were pulled. Caruso's team in New York examined all the operations to which Pitts had been assigned, along with all the files he had checked out. There was no smoking gun in the records. His carpool mate, who worked on a counterintelligence squad, said Pitts had never asked him for any documents. He passed a polygraph.

The results were a big zero. It appeared that if Pitts had been a spy, he'd gone dormant. The surveillances did not pick up the slightest hint that he was servicing dead drops or communicating with anyone suspicious.

Pickard conferred with Freeh, Assistant Director Bryant, his deputy, John Lewis, Mislock, and Wiser. They decided to run an undercover sting that would try to trick Pitts into reactivating. On August 26, they sent Dzheikiya to knock on Pitts's door in Spotsylvania, Virginia. "There is a guest visiting me. He wanted to see you," the Russian said. "He's in my car. He's from Moscow."

Across the street, FBI agents videotaping the encounter with a long lens saw sweat beads popping out on Pitts's face. An hour later, Pitts met Dzheikiya at Chancellorsville National Battlefield Park. Dzheikiya introduced him to a brawny man with a heavy Russian accent. He even held his cigarette the way Russians do, between the thumb and forefinger, but he worked for the CIA. Both he and Dzheikiya were wearing body wires, hidden listening devices. They gave Pitts a "tasking" letter and $15,000. "I'll do what I can," Pitts said.

Mary Pitts saw the look of panic on her husband's face when the Russian showed up at the door. Later that day, she searched her husband's home office and found the tasking letter and a pile of pornographic literature and a sex toy. She confronted her husband, who told her Dzheikiya was an old source from New York, and that the porn was "demonstration material" for his class on serial killers. His answers gnawed at her, so, three days later she called Tom Carter, an agent at the Fredericksburg, Virginia, FBI office, who had been the best man at Mary and Earl's wedding. She spilled out her fears and gave him a copy of the tasking letter. Then she went home, called her sisters and a neighbor, and told them what she had done. An FBI wiretap recorded her agony: "There is [sic] things wrong with this country but it's still my country. And passing information to a foreign national. . . . Could I have gone on with my regular and wonderful life? It's over, my life is over. It's over for me."

When Pitts came home, she told him about her call to Carter. Pitts calmly lied, saying the bureau was well aware of the Russian. The next day, he called Carter and gave him the same cover story he had given his wife.

But the FBI knew better, and it reeled Pitts in slowly, staging a series of undercover meets over the next sixteen months. In those meetings, Pitts gave the undercover officer information about his fellow FBI agents, suggesting who might be vulnerable to recruitment because of family or personal problems. As well, he handed over his FBI credentials — the holiest of holies, in bureau culture, plus a cipher lock combination, an FBI key, and an FBI scrambler phone. For his services he accepted $65,000.

One close call came on November 2, 1995, when Pitts, while hiding some money in the ceiling of his office at Quantico, discovered an FBI video camera. An FBI agent later slipped into the cubicle and labeled it as an "employee-attendance-monitoring device." That seems to have been enough, and Pitts continued to communicate with the undercover agent. The only sign he gave of concern was a letter he wrote to the "Russians," asking for $35,000 to $40,000 from "my account," an apparent reference to a KGB/SVR account in Russia. He said he wanted to use the money to fund an "escape" plan he intended to put in motion if he felt the FBI closing in on him.

In fact, the bureau was doing just that. In August 1996, Freeh promoted Pickard to special agent in charge of the Washington field office, so that he could supervise the investigation to its climax. Freeh, micromanaging again, ordered Pickard to have Pitts arrested on December 18. "It's gone on long enough," he said. "That's the day."

Pickard resisted. The agents were exhausted. He didn't want to disrupt their Christmas with court proceedings. Freeh had picked the date arbitrarily. Pitts wasn't going anywhere. "No, I want it done THAT day," Freeh ordered.

So, on December 18, 1996, a team of agents strode into room B-103 of the utilitarian FBI Academy in Quantico, Virginia, at 8:47 A.M. and handcuffed Earl Pitts. He was charged with taking more than $224,000 from the KGB between 1987 and 1992 plus another $65,000 from the FBI undercover operatives.

"These charges," an unsmiling Freeh told reporters later in the day, "are certainly repugnant to the fidelity, bravery, and integrity of the more than twenty-five thousand FBI employees, who day in and day out serve this great country, putting their lives and their welfare, and the welfare of their families, on the line."

Pitts pleaded guilty on February 28, 1997, and was sentenced to twenty-seven years in prison. Later that year, he gave a remarkably self-revealing interview to writer Marie Brenner, whose account of the case was published in the *Washington Post Magazine*. "If it were just a matter of greed, I would still be working for them," he told Brenner. The real answer, he said, was an inchoate rage that developed while he was working in New York. "I literally went through some kind of change that turned me inside out. Everything I held near and dear and important to me, I was ready to give up and trash. And that all happened in the space of three or four months.

"It's a strange thing," he added. "I would talk to a Russian, but not a psychiatrist."

Hanssen and Pitts had a lot in common. Hanssen had not gotten into a scuffle, verbal or physical, with any of his coworkers since the incident involving analyst Kimberly Lichtenberg in early 1993. Still, some who knew him sensed a bitter resentment that came out as caustic condescension. "I always felt terribly sorry for him because they didn't treat him very well," said one agent who saw him at FBI headquarters in the late eighties and early nineties. "He was very bright and introspective in a place that favors brashness and visceral responses. He was just smarter than other people, and he had a hard time not sounding superior, or like he was lecturing. As the years went by, he became less interested in trying."

One of those who did not appreciate Hanssen's lectures was his boss, Bear Bryant. Bryant had been at WFO at the time of Hanssen's assault on Lichtenberg and had had nothing to do with Bob's light disciplining. When Bryant took over the National Security Division toward the end of 1993, he heard the whole story and took an instant dislike to Hanssen. Bryant also chafed at the stinging comments Hanssen frequently made at meetings and felt that Hanssen was mocking him. Hanssen was a phony, too. He boasted to a college classmate in an e-mail that he had been chosen to debrief Ames. When Bryant learned of this claim much later, he laughed and said it was a lie. Nobody would have assigned Hanssen to a task that required people skills, he said. "He was such a dopey fucker," he said, "he never could have developed enough rapport to interview Ames."

In April 1994, Bryant saw that Hanssen was briefly reassigned to WFO, to work in the computer room. In December 1994, Hanssen returned to headquarters and was given the title of assistant to Bryant, which meant that there was no substantive job for him. By early 1995, Bryant had had enough of Hanssen. He arranged to transfer him out of the National Security Division — in fact, out of the Hoover Building altogether. A job had opened up at the State Department: the Office of Foreign Missions needed an FBI agent to work as liaison between State, the FBI, and the rest of the intelligence community. Bryant figured Hanssen would get along well enough with that crowd; and if he didn't, that was their problem. Bryant had his hands full with the aftermath of the Ames case and the incipient Pitts investigation.

Hanssen was packed off to the State Department on February 12, 1995.

16

THE STATE DEPARTMENT

Earl Pitts was debriefed through the spring and summer of 1997, and everything he said was factored into the matrix. Sometime around the middle of the year, the conclusion was unmistakable: there was another mole.

A very well placed one, too. Whoever had warned the KGB that Felix Bloch was under investigation was still out there. The matrix highlighted other compromises unexplained by the arrests of Ames, Pitts, and Harold Nicholson, a former CIA station chief in Romania. An electronic penetration was possible, but unlikely. Information coming in from the current crop of FBI and CIA sources inside Russian intelligence circles strongly suggested that the Russians had another American double agent. These sources did not know the mole's name nor where he worked. According to one official, "They had source reporting that there was a penetration in the CIA. Then, for three or four months, one source said it was in the FBI. Then the source came back and said it was in the CIA, definitely." If that were the case, it would make sense; the great majority of the blown operations occurred in the CIA.

There were two huge obstacles to overcome. First, the sources indicated the mole was dormant. Second, they said the KGB did not have his name. That seemed inconceivable. Wouldn't Moscow Center have insisted on knowing the name and position of the person who was selling such terrific material? "But if it was true, it made the job of identifying the person a thousand times more difficult," Mislock said.

The CIA Counterintelligence Center set a team to work compiling a long list of those with access to its compromised cases. The FBI did the same. Meanwhile, spy-hunters from both agencies pressed ahead with their recruiting drive, looking for ex-KGB officers who had answers.

The key was Viktor Ivanovich Cherkashin, who had been the KGB Line KR chief in the Soviet Embassy until the demise of the U.S.S.R. He had moved back to Moscow and was an official in the SVR. In *NIGHTMOVER,* his 1995 book about the Ames case, author David Wise reported that in 1986 Cherkashin received the Order of Lenin, the Soviet Union's second-highest honor. Wise attributed the decoration to Cherkashin's handling of Ames. But the FBI men believed Cherkashin was far more important than one case. "He was central to the whole issue of penetrations in New York and Washington," said Mislock. "There was a very intensive effort to account for every second of his existence." Now agents went into old Washington field office surveillance logs and case records, pulling together all that was known about Cherkashin's movements in the United States and his activities overseas. "Everything was put under the microscope to see if there was anything we missed," said Mislock.

The focus on Cherkashin turned up a great deal of interesting material, but it missed one of the most significant pieces of all. It was Viktor Cherkashin to whom Bob Hanssen had written on October 1, 1985, when he decided to resume spying for the Soviet Union. It was in that letter that Hanssen, using the alias "B," had named Martynov, Motorin, and Yuzhin as double agents. According to the KGB file, eventually obtained by the FBI, Cherkashin himself received and acted on the letters and packages delivered by "B"/Ramon Garcia. Cherkashin undoubtedly owed a big piece of his Order of Lenin medal to Bob Hanssen.

Bob Hanssen knew how important Cherkashin's silence was to his life. In 1993, Hanssen had met James Bamford, an ABC News consultant who specialized in spy stories. Bamford was an attractive guy, a muscular man of medium build, distinguished by his totally bald head and small, gray mustache. He was three years younger than Hanssen. The men were introduced by a mutual friend who worked for the CIA and thought they would have a lot in common.

Bamford recognized that Hanssen was a potentially good source for the book he was writing about the intelligence community. (Titled *Body of Secrets: The Anatomy of the Ultra-Secret National Security Agency,* it would later become a bestseller.) The two met on Bamford's sixty-foot motor yacht, *Safehouse,* moored on the Potomac River. "Our specialty was catching spies," Bamford said. "Mine from a journalist's point of view, his from the point of view of the FBI. We were just two colleagues. I had no problem talking about my work, and he was very interested." News reporters work hard to cultivate their sources, sometimes at social occasions, sometimes in the course of doing business. And government officials value news reporters who will take leaked information, publish it, and never name the source. In Washington, this trading of information is practiced as an art form.

Bamford liked Hanssen. The TV reporter's first impression was that, with his dark suit and pasty complexion, Hanssen looked the part of an FBI agent. "He had a gentle personality, an intellectual one," Bamford said. "He was very friendly, wouldn't interrupt, talked quietly. He came across as sincere." Quickly, the two men became friends. In December 1995, Bamford mentioned to Hanssen that he had flown to Moscow and landed a rare interview with Cherkashin (by then retired) about a tip he had gotten that there had been a KGB spy in the Carter White House. (The lead proved false.) "When I told Hanssen about the interview, he was amazed," Bamford said. "He told me that he was really important in the KGB. Bob was not very expressive, but I could tell he was surprised. He also asked me what we talked about." In many ways, the tools of a spy and those of a news reporter are similar. Listening is key. Most people make the novice's mistake of talking too much when they are trying to elicit information. Hanssen had a bad habit of lecturing people until they wrote him off as a pompous ass. And he certainly wasn't a pro, as his foolish approach to the GRU man demonstrated. But with Bamford, he handled himself well. He kept his mouth shut and just listened, smiling encouragingly every so often.

Over a few drinks, Hanssen and Bamford discussed Felix Bloch. Bamford, who had covered the investigation for *ABC World News Tonight with Peter Jennings,* had interviewed Bloch's girlfriend, Tina. He recounted details Tina had told about her sex life. "She was into S&M and went into much more than we could ever get on television news," Bamford said. Hanssen was interested in every detail — almost every detail, anyway. "He asked me all sorts of questions, but we never discussed sex," Bamford recalled. "He seemed very conservative, not the kind of guy I would talk about a *Playboy* article with."

In February 1995, when Hanssen was assigned to the State Department, he and Bamford continued to keep in touch, often for lunch. Occasionally, they ate in the State Department cafeteria, sometimes in an unpretentious restaurant a few blocks away called the West End Cafe. Usually they split the tab. "I like nice places to eat," Bamford said. "Bob wanted places where we could get a hamburger. He never expressed any interest whatsoever in money. He was on my boat and never said, 'Gee, I wish I had a boat.' I had a two-seater Mercedes, and he never said, 'Gee, I wish I had a sports car.'"

One subject that really got Hanssen excited was anticommunism. He repeatedly complained to Bamford about the ills of communism and how Russia was trying to infiltrate the United States. "For me, it went in one ear and out the other, but it was a big topic with him," Bamford said. "Bob would say that the Communists or the Chinese come in and pose as journalists. They will say they are doing a science piece for television on something like quantum computing. Then they would go to a university, get into a lab and start the camera rolling." Bamford said that in many ways, Hanssen was correct, that there were spies infiltrating the United States, and it was an interesting and legitimate topic: "But Bob would always fall into the red scare and talk about Soviet political philosophy."

In September 1995, Bamford invited Hanssen and his wife on his yacht, for a cruise up the Potomac River. Bamford also included the CIA intelligence officer who had introduced them, with his wife. The newspapers that weekend had been filed with stories about First Lady Hillary Clinton's speech at the United Nations' Fourth International Conference on Women, in Beijing. Mrs. Clinton was honorary chair of the event and gave a bold address that indirectly criticized China's coercive family-planning tactics. It was time "to break the silence" on human rights abuses against women, she said, to repeated cheers from her conference audience. The Hanssens were not fans of Bill or Hillary Clinton. Not only did they express disapproval of the First Lady's position, but they also went on to say that they were very opposed to the Planned Parenthood Federation of America. The intensity of their dislike for the family-planning organization surprised Bamford. "To me, Planned Parenthood was like the Boy Scouts or the PTA," he said.

Both Hanssens raised other conservative issues that day as well, including their opposition to abortion. "I got the feeling that supporting the antiabortion movement was one of their avocations," Bamford recalled. A year or two before, when he was still working at FBI headquarters, Hanssen told Bamford that he was taking a couple of hours of annual leave to attend an antiabortion rally. At the time, Bamford had not registered the depth of his friend's commitment. "This time, his main topic was antiabortion," Bamford recalled.

Most of all, Hanssen talked about religion. He wanted Bamford to know more about Opus Dei and tried to convince him to attend an Opus Dei meeting. "Week after week he brought it up," Bamford said. "He would try to get me back into the Church, tell me there were people there like myself, who had fallen away and come back. He would tell me the difference between Opus Dei and the Catholic Church." Bamford, who had been raised Catholic, says he did not want to hurt Hanssen's feelings by pleading disinterest: "I just told him I didn't like having to listen to all that Latin." But Hanssen was insistent. "He would say that Opus Dei is different, that the gatherings are more intimate and not as stuffy as a church," Bamford said.

Finally, Bamford agreed to accompany Hanssen to a meeting. They attended an evening event in an Opus Dei center near Tenley Circle, a facility later run by Louis Freeh's brother John, who, according to an Opus Dei spokesman, has since dropped out of the organization. In the Opus Dei tradition of separate meetings for men and women, that evening all the participants were male. A priest gave a short talk, and Hanssen introduced Bamford to a State Department colleague. "Bob was in his element," Bamford said. "He seemed well known and well liked. I had the feeling it was made up more of evangelical people than an ordinary church, that these people felt they were above the masses who only went to church."

One day Hanssen invited Bamford to accompany him to a gun show in Fredericksburg, Virginia, fifty miles south of Washington. Hanssen took his oldest son, Jack. "He had a lot of knowledge about guns," Bamford said of Bob. "He would pick one up and identify it." No one bought anything, perhaps because Hanssen didn't need anything; at home he kept a sizable collection of firearms, including an AK-47 automatic rifle, two Walther PPK pistols, a Colt Commando .38 special, a Colt Detective .38 special, a Smith and Wesson .38, a nine-millimeter Browning semiautomatic pistol, a Sig Sauer P228 pistol, a nine-millimeter Beretta, a Remington .308 rifle, and a Remington 870 shotgun. Few agents involved in counterintelligence work liked going to the range with their FBI-issue sidearms. But the idea of amassing a small private arsenal did not occur to them.

In April 1996, Bamford invited Bob and Bonnie Hanssen to his wedding. It was a relatively small affair, only about a hundred people, held at the National Press Club, in the heart of downtown Washington. "There were a lot of FBI agents there," Bamford said. "My best man was an FBI agent." But Bamford noticed something odd about Hanssen, a trait he had seen before but to which he had not given much thought. "Hanssen didn't socialize with them." Here he was at a festive occasion with FBI agents whom he had known for years, and he kept his distance. If anyone from the FBI took note, no one said a word.

*   *   *

Created by Congress in 1982, the Office of Foreign Missions (OFM) was Congress's way of attempting to regulate the activities of foreign envoys, who were often perceived as abusing their diplomatic immunity. The Office of Foreign Missions tracked parking and driving violations issued against vehicles under their control and applied sanctions when there were problems. The OFM also monitored all real estate transactions by foreign nations. For example, in June 2000, Xinhua, the Chinese news agency, attempted to buy a building in Arlington, Virginia, not far from the Pentagon. Hanssen was assigned to poll the intelligence agencies, who responded that the location would be an ideal place for the Chinese government to place antennae and scoop up Defense Department communications. "We turned the Chinese down," said Ted Strickler, deputy assistant secretary for OFM and Hanssen's boss at the time.

The State Department was a seven-story maze. Hanssen's office was located in room 2510-C, one of five small suites on the second floor, just around the corner from the main diplomatic lobby facing C Street. He arrived at the department every day about 8:15 A.M., sometimes after checking in first at the FBI to see if there were any new developments in his area.

To get to his office, Hanssen activated an access code on the main corridor and proceeded down a narrow, forty-foot-long hallway to a room painted the color of old eggshells. Then he walked through a series of two additional doors, each requiring different codes to be punched into the small, silver electronic cipher-locks. His office had a window, but the view wasn't much: industrial air conditioners on a roof in the courtyard. Hanssen's L-shaped desk held the computer that linked him to the FBI's main data system, for which he had top-secret access. To log on to the State Department's antiquated system, he had to leave his office and walk to another one, where he could gain access via an unclassified terminal.

The decorations in his office were sparse. His framed Knox College diploma and another from Northwestern University hung on one wall. The other wall displayed a picture of the Blessed Mother and a silver crucifix against a backdrop of burlap. (This is a practice of

Opus Dei followers, who often keep a small crucifix near their desk as an offering of one's work in gratitude for Christ's sacrifice on the cross.) While there were no family pictures around, Hanssen bragged about his kids, mentioning that Jack had started law school at the University of Notre Dame and that Susan, a student at Rice University in Texas, had joined Opus Dei.

Tom Burns had been Hanssen's supervisor in the FBI's Soviet analytical unit. When Burns moved to the State Department and became deputy assistant secretary in charge of OFM, he was Hanssen's boss again. Hanssen and Burns sometimes had lunch together in the State Department cafeteria. The conversation often turned to morality. "He was rabidly antiabortion," Burns said. "For him, there was no question that life began at conception. It came up any time we got into a discussion about values. I never heard him talk about sports."

John Lewis, who succeeded Bryant at FBI HQ as assistant director in charge of the National Security Division, ran into Hanssen a few times when he went over to the State Department for meetings. "He'd catch up to me in the elevators and he'd whisper, "Tell me how well I'm doing,'" said Lewis. "I think he was trying to get some kind of affirmation that he was doing well." Lewis, unlike his prickly predecessor, didn't dislike Hanssen, though he did find him strange. "Bob was a very quiet, reserved individual, who was exceedingly bright when it came to the use of computer and analytical studies, but he had a real problem relating to people," he said. "He was not your typical agent. Most agents did not like working with him, partly through his arrogance, partly through just his mannerisms. He kind of lived in a world of shadows. He'd come into a room and just stand there for a long period of time, not saying anything, anxious to try to please you." Lewis tried to find something kind to say, something that would make him feel more secure. But Hanssen was spending more and more time alone.

While Hanssen's communication skills were rudimentary at best, he was fluent in geek-speak. In many ways, it was a natural progression from his days putting ham radios together. On the Internet, nowhere as commonly used then as it is today, Bob had the ability to communicate with a tight circle of fellow nerds, who spoke the same code and helped each other through the confusing and frustrating challenges of trying to get their new programs to work. A techie in Australia communicating with someone in New York was far more interested in discussing the pros and cons of his hardware and software than the differences between kangaroos in the Outback and those in the Bronx Zoo. Unlike the State Department and the FBI where Bob was considered eccentric, this was a group where he was not only accepted, but appreciated. He belonged.

Hanssen became an Internet junkie, spending more and more time cruising cyberspace by himself. He did not have to be embarrassed by letting his emotions show because he wasn't communicating with anyone who knew him personally. This, too, was a mask he could wear. His colleagues at the office usually saw only sheepish, eager-to-please Bob Hanssen. But in his electronic correspondence to friends and colleagues, another side of Hanssen took center stage: the philosopher.

In one of several fascinating e-mail messages obtained by *Insight* magazine's Paul M. Rodriguez, Hanssen wrote: "You need to understand two things: 1) the real world; and 2) what it is like living with secret knowledge — withering, devastating insight which you can't tell anyone even when they make fun of you for holding such a ridiculous position." In another message, Hanssen wrote about the failure of the United States to recruit and maintain good spies: "All of our double-agent cases never make it out of the assumed controlled category because we can never make them convincingly productive. The KGB assumes they are bad and runs them for their productivity, such as it is, and to keep us busy with non-productive work. . . . They (the Russians) are right on this because they have gotten more good agents by that method than any other. Generally it's immediately obvious — a real agent grabs everything in his safe and brings it. Our DA's (double agents) bring one or two documents to each meeting. This is a big gulf. . . ."

Some of his messages were a bit nerdier. On November 11, 1995, Hanssen posted a message asking for advice on a technical issue: "I have recomplied the kernal (linux 1.2.13) with the sound drivers for my Proaudio 16. It complies and detects the card. Can someone help or suggest a new direction?" Another time, Hanssen asked for help on a topic called Missing Nameserver Packets: "I have the proper name-server address in my resolv.conf and nslookup and dsnquery work fine. Whenever I try to ftp or telnet to a regular alphabetic net address I get an unknown response. Running tcbdump shows the packets coming and going under dsnquery and nslookup." Should Bonnie have come upon him in the basement working on this stuff, as she had in 1979, she wouldn't have had a clue in the world what it was about. And that was just fine with Bob, who was growing more and more remote.*

About the time that the FBI was scrutinizing Viktor Cherkashin's contacts for clues to the identity of the double agent who blew the Felix Bloch case, Hanssen began worrying that the bureau had become sus-picious of his activities. It is not clear whether Hanssen learned that the PLAYACTOR matrix had spun off a new mole hunt after the arrest of Earl Pitts or whether Hanssen's guilt was starting to eat at him, but for whatever reason, Hanssen started querying the FBI's Automated Case Support System, which had come on-line in October 1995. Approved users, of which Hanssen was one, could access individual files by typing in a user identification number and a password. Once in the system, Hanssen could make full-text search requests for particular words or terms. Starting on July 25, 1997, Hanssen checked periodically to find out if his name, address, or the address of his drop sites were in the sys-tem and associated with the KGB or SVR. He also queried the system for indications that a Foreign Intelligence Surveillance Act (FISA) war-rant had been obtained for the cell phone in his name.

But at the same time as he was frantically checking the case support system, Hanssen had begun to act even more recklessly. When NATO

---

*New technology also offered another way for Hanssen to project himself as intelli-gent. For example, Ted Strickler, who would succeed Tom Burns as Hanssen's State Department supervisor in 1997, once asked Hanssen for advice about what cell phone to buy. "He said I should get Sprint PCS and explained the technicalities of how it worked, that the conversation was more private and less likely to be inter-cepted," said Strickler. The State Department official had no idea just how expert was the advice he received. It was the voice of authority, from a real pro, a guy who had, after all, a good track record for keeping conversations private.

escalated the bombing of the Yugoslav forces that were slaughtering ethnic Albanians in Kosovo, the Clinton White House expelled Yugoslav diplomats in Washington. Ted Strickler, Hanssen's boss at the State Department, was given the responsibility of issuing the formal deportation orders and searching for weapons and explosives within the Yugoslav Embassy. Hanssen was to walk with him to the door but had no other duties. As the midnight deadline approached, Hanssen let Strickler know that if there was any trouble, he — Bob Hanssen — was armed and ready. Strickler was flabbergasted; the spectacle of Bob Hanssen dropping a Serb diplomat or two in a wild gunfight was mind-boggling. If the Milosevic loyalists put up any resistance, it would be the job of the Secret Service to handle the situation with a minimum of lethal force. Strickler wrote Hanssen's behavior off to macho posturing, but in any case it had seemed odd, perhaps even irrational.

Hanssen took more intimate risks as well. Bob had always criticized his colleagues for their lewd fascinations. His own frequenting of the local strip club, and his relationship with Priscilla, showed that his high-mindedness was mostly a sham. Hanssen's Internet activities revealed that the strip club was only the beginning. On June 5, 1998, Hanssen went into a newsgroup on the web called "alt.sex.stories" and posted a scene that seems to have come from the earliest days of his marriage. In an extraordinary breach of security regulations, not to mention a giant lapse of common sense, Hanssen used his real name, his own computer, and his real e-mail address:*

From: Robert P. Hanssen (hanssen@nova.org)
Subject: Bonnie (wife, exhib, true)

*Would it be possible for someone with details of Hanssen's life to post such a scene on the Internet, dated June 5, 1998, using Hanssen's name and e-mail address *after* Hanssen's arrest? David Krane, a press spokesman for Google, an Internet search engine that purchased and maintained the Usenet archives, said that could not happen. He said that because entries are time-stamped when they are received, it was impossible for anyone to submit this item as a hoax. "There is no way to get the date in earlier," Krane said.

Newsgroups: alt.sex.stories
(This is the only article in this thread)
Date: 1998/06/05

It was only around four in the afternoon, and Bonnie still had plenty of time as she walked over and perched on the high wooden stool. She sat, freshly showered and still naked, in the warm light of the summer Chicago sun which streamed through her apartment's large bedroom window to her left. Refreshed from her shower after teaching second grade at the parish school, it was time to fix her hair. This was her habit, her little ritual after a shower, a time to herself to unwind and feel feminine, a time to feel the air on her skin and fix herself all pretty for Bob.

Hanssen went on, detailing Bonnie's insecurities about her looks (her breasts looking like they were "about to pop out of her sundresses") before turning back to Bob.

Bob was a dental student. He'd just started the clinical portion of the program. He'd scored in the top two percent on his national Dental Boards, and Bonnie was proud of him. She was going to show him a good time. Bonnie knew that a good time meant letting Bob show her off. Bob loved having men's tongues dangle out looking at his wife. So tonight Bonnie had vowed to herself to make that happen.

Bob was a leg man, Bonnie knew that, and Bonnie knew she had legs that could handle that. She'd learned shopping with him that no dress was too short. So, tonight Bonnie intended to do something she'd never allowed herself before, to push the limit in that direction — to please him. She'd hunted for and found a secret weapon — a dress, a special one for a special occasion. She'd found it in a store down on Rush Street. Bonnie was quite innocent and naive in many ways really, and had no idea it was a store catering to strippers. Bonnie hardly even knew about strippers. She'd just been out shopping and happened upon it and gone in. She was just amazed that a store carried so many sexy dresses.

After some more about Bonnie's short dress, Hanssen — strippers on the mind — put his wife on stage:

Bonnie was playing with her hair, trying it different ways. She tried it up, she tried it down, and was about to settle for up when she noticed an odd movement out on the elevated train tracks across the alley from her bedroom window. She looked out the window quickly. "My God!" she thought. There were five workers standing leaning on their shovels looking at her. In a panic, she bounded from her stool across the bed to try to grab the shade and pull it down. Because the bed stood only a foot below the window and along the wall, this move necessitated her standing stark naked on the bed to reach up for the shade pull. She was there in full view of her suddenly bemused audience. Bonnie grasped the shade and pulled it down, with short-lived relief. The shade didn't catch and flapped up again. She sprung back a second time, her cute little bush fully exposed, and tried again. She yanked it down again and again it flapped open. Worse, this time instead the cord tied itself around the shade roller. Bonnie went up for it again. Bonnie's face was flushed. The men were looking right at her and she was totally naked. It seemed like forever while Bonnie stood in that window trying to untie it, but she got it. This time, a little calmer from the delay, she laughed at the smiles of her audience and closed it slowly and deliberately like putting the curtain down for them on a good show, and had even given them a little wave goodbye.

. Then Bonnie collapsed panting from the excitement on her bed. Her heart was pounding. She felt galvanized as if by electricity from the experience. She realized she felt something else too. She felt aroused. "If only Bob were here," she thought, I'd show him even a better time than the workers on those tracks.

The rest of the posting described Bonnie showing herself to the workers, becoming increasingly aroused, and then gearing up for Bob's arrival.

In her heels, Bonnie bent over to look at herself from the rear. She thought, "Opps." [sic] She told herself she would have to remember not to bend over like that, but then removed her panties too. Bob was going to get his money's worth tonight.

The Internet posting described real details of Hanssen's days as a young married man: that he was a dental student and "leg man," married to a pretty, Catholic woman named Bonnie, whom he met in a mental hospital where they both worked, and that they lived in an apartment on Winthrop Avenue in Chicago. Whether or not the actual scene ever took place, such a public posting would be enormously embarrassing for any woman, especially the mother of six children, who was teaching religion at a private girls' Catholic school.

An even more explicit account discussing Bonnie and Jack Hoschouer was posted on an Internet Usenet adult site on March 7, 1999:

Date: 1999/03/07

The "Unwitting?" Porn Star (mmf, exhibitionism)

For years, I have sent nude pictures of Bonnie, my wife, to my close friend Jack. It began back in the Vietnam War years. Jack was over there and I was safe here at home. I'd tried to join the Navy as a dental student, but couldn't see well enough — too nearsighted, even though I was corrected with contact lenses to 20:20. That same problem kept me out of the draft. I felt bad. Here Jack was making all these sacrifices for my country and there was so little I could do to support him. Well, Bonnie was just 21 then and we'd just married. Jack had been my best man, and Bonnie was very pretty in those days, still is of course though she is older now. She was only 5'4" but shaped 34–23–36 with long brown hair and big brown doe eyes. Gorgeous would not be an overstatement. Pretty as she was, she made an excellent subject for photography. She tolerated posing nude for her husband too. Actually, she claimed to tolerate it, but when she got going posing nude she was like a spaniel to water. She put her soul into it, and the resulting pictures were electrifying. Now, my hobby was always photography and Jack's was too. Even in our little apartment while I was in school, I had set up the bathroom with a board I could put on the tub to hold the enlarger above the trays and the waterbath below. It worked well, and I had this great supply of knockout pictures of Bonnie. Unfortunately, I had no one to whom I could appropriately show them. She wasn't about to let me show them either, artistic though they were! So one day, I sent a few off to Jack in one of my regular letters — he loved it. He got to see a whole side of Bonnie he never knew existed. Technically, I guess you could say he got to see all sides, crevices, and cracks of Bonnie, but that might be a bit crude. (Bonnie did a lot of posing with her legs spread. It turned her on. Bonnie was a great one for generating photos of "artistically posed girl with wet and gaping pussy." They all went to Jack.) . . .

When I finally bought a house, I went to work and built a darkroom down in the basement and stored all of Bonnie's photos there. Bonnie was having problems getting pregnant then so there were no kids around to get into things like that. In the end, Jack came back from the war with a Bronze Star and other decorations to make a career as a military officer, visiting often when blowing through town enroute to some military assignment. Bonnie enjoyed him greatly and was as proud as I was of the risks he'd undertaken for his country. He'd survived where others hadn't.

Bonnie even posed for Jack occasionally, including some in a tight sweater that sizzled, but never nude, probably because Jack could never get up the guts to ask her. She did offer, on her own however, to let him come to our house any time he wanted to use the darkroom, knowing her pictures were stored there. . . . When I found out she'd let him use the darkroom, I thought the time was ripe to let her know that Jack had seen the nudes of her. I wanted her to know he'd already seen her so she'd have nothing to be shy about in case she was ever tempted to pose for him, and this way I could blame it on her. So when she told me about it, I said, "Oh Bonnie, all your pictures were down there, and Jack told me he saw them." I added, "and thought they were great." She was greatly embarrassed, said she'd forgotten and all, but I was never completely convinced. Jack had often dreamed of photographing Bonnie himself or of just seeing her nude in person. One never knows about the photography. It may come to pass, but we could do something about the seeing. That could be arranged. We initially made some forays in this regard where Jack watched Bonnie shower through the glass shower-stall door. I would talk to her as she showered, and he would look through the bathroom doorway from the darkened bedroom behind me. This was fun but not totally satisfying. Our schemes progressed in this regard to the point of letting Jack see Bonnie and I having sex. The first time we did this Jack was visiting for a week. It was a warm fall. Each night, I'd leave the shade up a bit and leave our bedroom window open about six inches so Jack could come up and stand on a pre-positioned chair on our deck and look in while I had sex with her at night. That worked like a charm. He could see her walking around the room naked and I'd position her in different ways on the bed while fucking her so he'd get a good look of my cock going in and out or of her tits bouncing. By pure chance, to his good fortune, she even bent over right in front of the window once when he was there, and he got a good view of her pussy from about a foot away. It was great. I was dying watching. Our house backs on the woods so there was really little risk to him in doing this, but it still got old because of the chill and risk. It was then that technology came to the rescue. At a security show, we bought subminiature video cameras designed for surveillance purposes. We bought two. One was no more than a little one-inch by one-quarter inch box with a low light CCD sensor that looks through a pinhole. The other was slightly larger; a high quality but very small lensed camera. We also bought video transmitter/receiver combinations, which could relay the signals from the cameras in place in our bedroom down to our den. After some experimentation (while Bonnie was out of town for a few days,) we set up a great system. Jack could sit in our den when he visited and see everything up in the bedroom. This proved no end of fun during his visits.

Often now he'll stay five or six days doing things like research on his Ph.D. thesis at the various libraries here. On recent trips, in the mornings he sits and watches Bonnie on the large screen TV in our den as she gets showered and dressed. (Bonnie still fixes her hair while nude each morning.) At night, he watches the nightly sex scene or, once, Bonnie modeling her Victoria-Secret white nylons, sheer bra, and heels for me before we fucked. That really got us all going.

This whole business has been no end of fun. He can even tape the sessions on our VCR. Bonnie may be the only teacher at the elite girl's school where she works who is also a porn star! Of course, she doesn't know it. Well at least we think she doesn't know it. I do notice that when Jack visits she wants sex every night. Perhaps it is just that I'm always ready for it when he's here as you can imagine. Of course, Jack flatters her a lot too. He's still hoping some day for THE BIG modeling session, and women love to be flattered.

Anyway, Jack and I have our fun. Bonnie looks great. Jack and I love seeing her tits slapping together as she takes cock hard. She is now size 36 on top and she's kept her waist and hips trim. She's still a beautiful woman.

Individuals close to Hanssen confirm that over the years, he sent nude pictures of his wife to Jack Hoschouer, and that Hoschouer protested in writing. Yet the practice continued. Repeatedly asked for comment on the Internet posting, Hoschouer replied only, "Bob liked to fantasize."

At the same time that Hanssen was posting sexual fantasies on the Internet, he was giving friends the impression that he had become more religious. "In the last three-to-four years, we had intense theological discussions by e-mail," Hoschouer said in a conversation before being questioned about the Internet postings. "He was accepting the real orthodox line to the point where he never would question anything the Church said." Hoschouer grew tired of the litany. "Bob told me at one point that he didn't believe in evolution," Hoschouer said, "that everything was created at the same time and some species died out." Hoschouer couldn't resist goading Hanssen a bit. "That doesn't explain why whales have rudimentary hip bones," he shot back. Hanssen's reaction was immediate. "Bob got really quiet," Hoschouer said. "With Bonnie, you could argue. Bob got angry. I seldom discussed these things face-to-face because he got so mad. It was easier in e-mail because he could blow up alone."

A peeping tom — that was what seemed to turn Bob Hanssen on. It made some sense. Hadn't he always been the outsider, looking in? The prowler who had been at the window? And wasn't that what being powerful was all about? Knowing somebody's secrets? Bob Hanssen may not have believed in evolution, but *he* certainly had changed over the years. Once, perhaps, there had been one Robert Hanssen, but now there were many of them. The devout Catholic who wrote sick porn about his wife. The loyal husband who mailed out nude photos of his beloved without her knowledge. The anti-communist who got a paycheck from men whose job it was to support the Stalinist cause. The American patriot who was an American traitor. And he was about to go back to work.

17

SPYING RESUMES

Sometime in 1998, the matrix produced a short list of individuals who could have blown the Felix Bloch case. One in particular seemed a good candidate: a CIA intelligence officer in his fifties who had spent much of his sixteen-year career at the agency in counter-espionage. A former U.S. Air Force intelligence officer, the CIA man had specialized in the difficult task of ferreting out "illegals," deep-cover KGB officers who did not have diplomatic status. He had been decorated for his extraordinary work unmasking Reino Gikman, a KGB illegal based in Austria. It had been that breakthrough that led the CIA to intercept Gikman's April 27, 1989, call to State Department diplomat Felix Bloch in Washington.

The mole hunters had other reasons to single out this CIA man for further investigation: he had access to a number of other operations believed to have been compromised. "The evidence we had at the time pointed ninety percent to the problem that had been identified in the agency," said one FBI official involved in the investigation. The CIA officials involved concurred that the man was a good candidate because of his access, and that details of his life fit the description of a double agent who had been labeled "B" in KGB files. By now, the CIA and FBI had developed a number of current and former Russian intelligence officers as sources. These reported that "B" had used Nottoway Park and other northern Virginia parks as a dead drop. The CIA officer lived near Nottoway Park and jogged there frequently. His home was, in fact, very close to Bob and Bonnie Hanssen's old house on Whitecedar Court.

The man was given a new CIA job that reduced his access to sensitive information. A team of FBI agents working in the Washington field office's National Security Division and investigators from the CIA Counterintelligence Center came up with a couple of ruses to test him. First, CIA officials told him there had been a major breakthrough in the Bloch case, that a Russian defector was on the way to the States with the key to the identity of the person who had warned off Bloch. When he was asked to join the debriefing team, his reaction was gauged, and he was given a polygraph test, ostensibly a requirement for participating in the case. After he passed, he was informed the defector wasn't coming after all.

In November 1998, a man knocked on his door and, speaking in an accent, warned him that the CIA was on to him. The man handed him an "escape plan" and instructed him to go to a subway station in northern Virginia the next night. This gambit was designed to reactivate him if he was dormant, but he didn't bite. Instead, he reported the approach to the FBI. He even helped a sketch artist come up with a portrait of the stranger.

In February 1999, Tim Caruso, the father of the matrix and the driving force behind the Pitts case, was promoted from Eurasian section chief at FBI headquarters to special agent in charge of the National Security Division at the Washington field office, replacing Sheila Horan, who moved to headquarters. Under Caruso's leadership, the mole hunt was moved from covert, "non-alerting" status, to overt, meaning that the FBI agents felt they could do no more to resolve the matter without conducting interviews of which the suspect would become aware. That month, the FBI agents interviewed a female CIA employee who had worked on the Bloch case with the suspect. According to John Moustakas, a lawyer engaged by the male CIA officer, the agents indicated to the woman that her friend was under investigation for espionage. They questioned her about parks in northern Virginia, about trips to New York, about his interest in diamonds. They asked her if she had ever helped the man place an ad in the *Washington Times*. All of her answers were negative — she had seen nothing unusual in her friend's behavior or tastes.

Though the mole hunters still had no substantive evidence against the CIA officer, they persisted in the belief that he was their man. This led Caruso's team of agents to engage in tactics that would prove highly controversial. On August 18, 1999, they confronted the man and his family with their suspicions. The only evidence the bureau had against him was circumstantial, mainly the information from the matrix about his access to blown cases and a couple of maps of Nottoway Park with numbers jotted on them, found in a covert search of his house.

When the FBI agents showed the suspect one of the maps, he replied, "That's my jogging map," explaining that the CIA officer had drawn it himself; the numbers were his usual times to reach each landmark. He denied being a spy. While he was being interviewed, the FBI team attempted to find holes in his story by sending agents to confront his three adult children, his ex-wife, and his two sisters.

The man's daughter, who worked in the personnel office of the CIA, would later tell the *Washington Post* that she was called into a room and met by an FBI agent who said, "Please sit down. We have some bad news for you. Your father is a spy."

"They told her he was guilty of a capital offense and would soon be arrested," said Moustakas, who denounced the agents' statements to his family as "outrageous."

Senior FBI officials later denied that the agents bluffed with an assertion that they had hard proof of the father's guilt and were on the verge of arresting him; the agents had been instructed to say only that he was "under investigation for espionage" and leave it at that. FBI officials acknowledged, however, that the man's children, presented with this frightening message, might well have become alarmed and assumed that their father's arrest was imminent.

Whatever the case, both sides agree that the daughter and her two brothers corroborated their father's claim that the Nottoway Park sketch was nothing more than a jogging map; also, that the children seconded their father's assertion that his computer skills weren't remotely sufficient to have enabled him to encrypt messages to the KGB on diskettes. Moustakas said the man asked to take a polygraph test to clear his name but was rebuffed. "They had nothing but whispers of evidence," Moustakas added. "They were desperate to make the case. There were no financial spikes in his record. They had learned nothing through electronic and physical surveillance. He passed their sting operation test. At some point, you can't keep rejecting exculpatory evidence."

Still suspicious, the CIA took away the suspect's security clearance and put him on administrative leave. In October 1999, a Bell Atlantic repairman, called to his home because his telephone was on the blink, announced, "This seems to be your trouble," and fished a tiny bugging device out of the phone.

By this time, many of the man's colleagues knew about the investigation because they had been personally questioned by the FBI. Did Bob Hanssen know, too? The answer is not clear. But within weeks of the CIA man's interrogation, Hanssen wrote to an SVR Line KR officer and expressed his desire to reactivate "Ramon Garcia" as a spy. The Russians were delighted. They had not forgotten about Ramon, as an effusive letter from the SVR indicated:

Dear friend: welcome!

It's good to know you are here. Acknowledging your letter to V.K. we express our sincere joy on the occasion of resumption of contact with you. We firmly guarantee you for a necessary financial help. Note, please, that since our last contact a sum set aside for you has risen and presents now about 800.000 dollars. This time you will find in a package 50.000 dollars. Now it is up to you to give a secure explanation of it. As to communication plan, we may have need of some time to work out a secure and reliable one. This why we suggest to carry on the 13th of November at the same drop which you have proposed in your letter to V.K. We shall be ready to retrieve your package from DD since 20:00 to 21:00 hours on the 12th of November after we would read you [sic] signal (a vertical mark of white adhesive tape of 6–8 cm length) on the post closest to Wolftrap Creek of the "Foxstone Park" sign. We shall fill our package in and make up our signal (a horizontal mark of white adhesive tape). After you will clear the drop don't forget to remove our tape that will mean for us — exchange is over.

We propose a new place where you can put a signal for us when in need of an urgent DD operation. LOCATION: the closest to Whithaven [sic] Parkway wooden electricity utility pole at the south-west corner of T-shaped intersection of Foxhall Road and Whitehaven Parkway (map of Washington, DC, page 9, grid B11). At any working day put a white thumb tack (1 cm in diameter, colored sets are sold at CVS) into the Northern side of the pole at the height of about 1.2 yards. The tack must be seen from a car going down Foxhall Road. This will mean for us that we shall retrieve your package from the DD Foxstone Park at the evening of the nex [sic] week's Tuesday (when it's getting dark).

In case of a threatening situation of any kind put a yellow tack at the same place. This will mean that we shall refrain from any communication with you until further notice from your side (the white tack). We also propose for your consideration a new DD site "Lewis". DD LOCATION: wooden podium in the amphitheatre of Long-branch Nature Center (map of N.Virginia, page 16, grid G8). The package should be put under the FAR-LEFT corner of the podium (when facing the podium). Entter [sic] Longbranch Nature Center at the sign from Carlin Springs Road (near 6th Road south) and after parking your car in the lot follow the sign "To Amphitheatre." LOCATION OF THE DD SIGNAL: a wooden electricity utility pole at the north-west corner of the intersection of 3d Street and Carlin Springs Road neaqr [sic] the Metrobus stop (the same map, grid F7). The signals are the same as in the "Foxstone Park" DD. The white adhesive tape should be placed on the NORTHERN side of the pole, so that it could be noticed fro [sic] a car moving along Carlin Springs Road in the southern direction from Route 50. Please, let us know during the November operation of your opinion on the proposed places (the new signal and DD "Lewis").

We are intending to pass you a permanent communications plan using drops you know as well a new portion of money. For our part we are very interested to get from you any information about possible actions which may threaten us.

Thank you. Good luck to you.

Sincerely,

Your friends

Why did he take the risk of reactivating? There seemed little chance that the FBI would catch up with his past espionage as long as he stayed dormant. And by all indications, its attention was focused solely on the now-disgraced CIA man. But by the fall of 1999, the family's college and private school tuition bills were putting a major crunch on the Hanssen family budget. Mark was a freshman at the University of Dallas; Jack, a senior at the College of William and Mary. Greg was still at the Heights, and Lisa was planning to attend Oakcrest, where Bonnie was working as a part-time religion teacher. All of the children had generous scholarships and student loans, but juggling expenses was becoming more and more difficult. To meet the bills, the Hanssens had refinanced their house several times.

The strain showed in Bob's e-mails to Jack Hoschouer. Over the past eight to ten years, Jack had found Hanssen increasingly rigid and doctrinaire. But now it was much worse. The two friends had spent their lives debating controversial issues, even politics, but now Jack found Bob unpredictable and quick to anger. Their discussions became especially difficult when the subject turned to religion, Hanssen insisting that Hoschouer accept the Church's view on whatever matter they were discussing. "He would say, 'The Church says so, so it's true,'" Hoschouer explained. "I decided to hang myself out. I said, 'You don't realize that the Church's method of arriving at the truth is the same process used by the Central Committee of the Soviet Union. There is a small body with the truth, and they debate it, and the result is party line.' Bob said, 'Of course, this is true, but one system stood for evil and one for absolute good.'" Finally, the two men stopped discussing the subject of religion face-to-face, but even apart from the debates, Hoschouer recalled that Hanssen's electronic messages were growing bizarre: "He would e-mail me: 'Do Not Be Deceived.'"

At one point, Hanssen brought up the case of Jonathan Pollard, the American who was convicted of spying for Israel in 1985. "Bob's attitude was a spy's a spy's a spy," Hoschouer said. "Whoever it is, they are unlawfully disclosing their country's secrets." When the topic switched to John Walker, the retired sailor convicted of spying for the Soviets, Hoschouer wrote: "We should take Walker out and shoot him." Bob agreed.

At another point, Hanssen sent Jack a gift, a new Rolex watch. Jack noticed a similar one in a store and the price was $3,475. "Maybe it was worth sixteen hundred dollars when he gave it to me," Jack said. He didn't ask Hanssen where the watch came from, figuring it was just one of those things that sometimes fall into the lap of traveling bureaucrats, and that reporting it to the government, which is, of course, the law, was conveniently overlooked.

In retrospect, it seems that Hanssen was continuing to drop clues, almost daring people to catch him, or at least to pay some sort of attention. About a year before the arrest, Hanssen and Hoschouer got into an e-mail discussion about penitence. "Bob said penitence should be associated with the penitentiary," Hoschouer recalled. "You go in the slammer and repent and resolve to be better."

There was at least one aspect of Hanssen's personality that never varied. A decade after the Cold War ended, he was still lecturing anyone who would listen about the evils of communism. He had written a long government report titled *Soviet Active Measures in the Post–Cold War World,* a dense, detailed analysis of Russian politics between 1988 and 1991. Hoschouer said Hanssen was extremely proud of the report, which he later turned into a speech that he would give to civic groups and Opus Dei meetings.

Meanwhile, Hanssen once again went looking for love, or something like it, in the wrong places. He resumed his visits to Joanna's, this time fixating on a slim Vietnamese stripper who danced under the name of "Brooke." He gave her a tip, and "she came over and started to talk," recalled another patron. It didn't stop there. Hanssen went to see Brooke dance a second time, before apparently concluding that she was beneath him. He later wrote an e-mail to a friend, detailing Brooke's "bad physical points." There he was, again standing in judgment, this time of a stripper's looks.

In his third life, as Ramon Garcia, secret agent, Hanssen spent a lot of time on the FBI's Automated Case Support System (ACS). Using either that terminal or one within the National Security Division, he queried the system on October 21, 26, and 27, November 3, and November 15 for keywords: DEAD DROP, VIENNA, FCI, FOX-STONE, VIRGINIA.

On January 13 and 18, 2000, he checked the terms: DEAD DROP, DROP SITE AND VIRGINIA, SVR AND DEAD DROP NOT GRU. On March 14, he checked DEAD DROP AND SVR.

But Hanssen had lost his old confidence. Ramon's tone was no longer clipped and commanding. On or before March 14, Hanssen wrote a self-revelatory letter to the SVR in which he intimated he was spinning out of control. The letter read, in part:

> . . . I have come about as close as I ever want to come to sacrificing myself to help you, and I get silence. I hate silence. . . .
>
> Conclusion: One might propose that I am either insanely brave or quite insane. I'd answer neither. I'd say, insanely loyal. Take your pick. There is insanity in all the answers.
>
> I have, however, come as close to the edge as I can without being truly insane. My security concerns have proven reality-based. I'd say, pin your hopes on 'insanely loyal' and go for it. Only I can lose. I decided on this course when I was 14 years old. I'd read Philby's book. Now that is insane, eh! My only hesitations were my security concerns under uncertainty. I hate uncertainty. So far I have judged the edge correctly. Give me credit for that. Set the signal at my site any Tuesday

evening. I will read your answer. Please, at least say goodbye. It's been a long time my dear friends, a long and lonely time.

    Ramon Garcia

Hanssen used the word "insane," or variations of it, a half-dozen times here, and this was not the only letter in which he used it. The *Washington Times* later published on its Web site an e-mail he had sent to a colleague in March 1999, which he began, apropos of nothing, "The problem with genius is that it often borders on insanity. The problem with truth is that it sometimes seems utterly fantastic."

For some reason, Hanssen still felt he had to impress his handlers and fell into his old habit of exaggerating even when the Russians, of all people, would know better. The notorious spy Kim Philby wrote his book, *My Silent War*, in 1968, when Hanssen was twenty-four — the year he and Bonnie were married. Hanssen was an unhappy dental school student still trying to figure out what direction his life should take. He had been nineteen and a college student when news of Philby's spying broke. If Hanssen were serious that as a young man he admired Kim Philby, it suggests he had been thinking about the possibility of spying for a very long time. But there was no way he had been inspired by Philby at the age he claimed.

On March 31, Hanssen queried the FBI computer for DEAD DROP AND RUSSIA, and on May 22 he entered his address, TALISMAN DRIVE. The coast looked clear, and, thus emboldened, Ramon wrote the SVR again, around June 8.

Dear Friends:
    Administrative Issues:
    Enclosed, once again, is my rudimentary cipher.
    Obviously it is weak in the manner I used it last — reusing key on multiple messages, but I wanted to give you a chance if you had lost the algorythm [sic]. Thank you for your note. It brought me great joy to see the signal at last. As you implied and I have said, we do need a better form of secure communication — faster. In this vein, I propose (without being attached to it) the following:

One of the commercial products currently available is the Palm VII organizer. I have a Palm III, which is actually a fairly capable computer. The VII version comes with wireless internet capability built in. It can allow the rapid transmission of encrypted messages, which if used on an infrequent basis, could be quite effective in preventing confusions if the existance [sic] of the accounts could be appropriately hidden as well as the existance [sic] of the devices themselves. Such a device might even serve for rapid transmittal of substantial material in digital form. Your FAPSI could review what would be needed, its advisability, etc., obviously — particularly safe rules of use. While they can be quite effective, in juggernaut fashion, that is to say thorough. . . .

New topics:

If you are wise, you will reign [sic] in the GRU. They are causing no end of grief. But for the large number of double-agents they run, there would be almost no ability to cite activity warranting current foreign counterintelligence outlays. Of course the Gusev affair didn't help you any. If I'd had better communications I could have prevented that. I was aware of the fact that microphones had been detected at the State Department. (Such matters are why I need rapid communications. It can save you much grief.) Many such things are closely held, but that closeness fails when the need for action comes. Then the compartments grow of necessity. I had knowledge weeks before of the existence of devices, but not the country placing them. . . . I only found out the gruesome details too late to warn you through available means including the colored stick-pin call. (Which by the way I doubted would work because of your ominous silence.) Very frustrating. This is one reason I say "you waste me" in the note. . . .

The U.S. can be errantly likened to a powerfully built but retarded child, potentially dangerous, but young, immature and easily manipulated. But don't be fooled by that appearance. It is also one which can turn ingenius [sic] quickly, like an idiot savant, once convinced of a goal. The purple-pissing Japanese (to quote General Patten [sic] once again) learned this to their dismay. . . .

I will not be able to clear TOM on the first back-up date so don't be surprised if we default to that and you find this then. Just place yours again the following week, same protocol. I greatly appreciate your highly professional inclusion of old references to things known to you in messages resulting from the mail interaction to assure me that the channel remains unpirated. This is not lost on me. On Swiss money laudering [sic], you and I both know it is possible but not simple. And we do both know that money is not really "put away for you" except in some vague accounting sense. Never patronize at this level.

It offends me, but then you are easily forgiven. But perhaps I shouldn't tease you. It just gets me in trouble.

thank you again,
Ramon

His SVR handlers wrote back around July 31.

Dear Ramon:

We are glad to use this possibility to thank You for Your striving for going on contact with us. We received Your message. The truth is that we expended a lot of efforts to decipher it. First of all we would like to emphasize that all well known events wich [sic] had taken place in this country and in our homeland had not affected our resources and we reaffirm our strong intentions to maintain and ensure safely our long-term cooperation with You.

We perceive Your actions as a manifestation of Your confidence in our service and from our part we assure You that we shall take all necessary measures to ensure Your personal security as much as possible. Just because proceeding from our golden rule — to ensure Your personal security in the first place — we have proposed to carry out our next exchange operation at the place which had been used in last august [sic]. We did not like to give You any occasion to charge us with an inadequate attention to problems of Your security. We are happy that, according to the version You have proposed in Your last letter, our suggestions about DD, known as "Ellis", coincided completely. However a situation around our collegues [sic] at the end of passed [sic] year made us to refuse this operation at set day.

1. We thank You for information, wnich [sic] is of a great interest for us and highly evaluated in our service.

We hope that during future exchanges we shall receive Your materials, which will deal with a work of IC [the intelligence community], the FBI and CIA in the first place, against our representatives and officers. We do mean its human, electronic and technical penetrations in our residencies here and in other countries. We are very interested in getting of the objective information on the work of a special group which serches [sic] "mole" in CIA and FBI. We need this information especially to take necessary additional steps to ensure Your personal security. . . .

2. Before stating a communication plan that we propose for a next future, we would like to precise [sic] a following problem. Do You have any possibility to meet our collegues [sic] or to undertake the exchange ops in other countries? If yes, what are these countries? Until we receive Your answer at this [sic] questions and set up a new communication plan, we propose to use for the exchange ops DD according to the following schedule:

= DD "LEWIS" on 27 of may 2001 (with a coefficient it will mean on 21 of november 2000). We draw Your attention on the fact that we used a former coefficient –6 (sender adds, addressee subtracts). A time will be shown at real sense. We will be ready to withdraw Your package beginning by 8 PM on 27 may 2001 after we shall read Your signal. After that we put DD our package for You. Remove Your signal and place our signal by 9 PM of the same day. After that You will withdraw our package and remove our signal. That will mean an exchange operation is over. We shall check signal site (i.e., its absence) the next day (28 of May) till 9 PM. If by this time a signal had not been removed we shall withdraw our package and shall put it in for You repeatedly dates with DD "ELLIS" — in each seven days after 28 May till 19 of June 2001 (i.e., 13 of December 2000).

= We propose to carry out our next operation on 16 of october 2001 (i.e., 10 of April) at the DD "LINDA" in "Round Tree park" (if this place suits for Your [sic] we would like to receive Your oppinion [sic] about that during exchange in may). A time of operation from 8 pm to 9 pm, signals and schedule of alternate dates are the same. In the course of exchange ops we shall pass to You descriptions of new DD and SS that You can check them before. You will find with this letter descriptions of two new DD "LINDA" and "TOM". Hope to have Your opinion about them.

In case of break off in our contacts we propose to use DD "ELLIS", that you indicated in your first message. Your note about a second bridge across the street from the 'F' sign, as back up, is approved. We propose to use "ELLIS" once a year on 12 August (i.e., with coeff. it will be 18 February) at the same time as it was in August 1999. On that day we can carry out a full exchange operation — You will enload your package and put a signal, we shall withdraw it, load our package and put our signal. You will remove our package and put your signal. Alternate dates — in seven days 'til next month.

= As it appears from your message, you continue to use post channel as a means of communication with us. You know very well our negative attitude toward this method. However if you send by post a short note where date (i.e., with coefficient), time and name of DD for urgent exchange are mentioned, you could do it by using address you had used in September (i.e., with coeff.) putting in a sealed envelope for V.K. In future it is inexpedient to use a V.K. name as a sender. It will be better to choose any well known name in this country as you did it before.

3. We shall continue work up [sic] new variants of exchanging messages including PC disks. Of course we shall submit them to your approval in advance. If you use a PC disk for next time, please give us key numbers and program you have used.

4. We would like to tell you that an insignificant number of persons know about you, your information and our relationship.

5. We assess as very risky to transfer money in Zurich because now it is impossible to hide its origin. . . .

While his devotion to the SVR seemed solid, Hanssen was starting to think about life after the FBI. On April 18, 2001, he would be fifty-seven and required to hand in his badge. Most retired FBI agents were double-dippers, working in private industry while collecting and investing their FBI pensions. If they were clever about it, when they hit sixty-five, they could look forward to corporate retirement benefits as well as a government pension. With this in mind, Hanssen composed and sent a résumé to a friend in a nationwide accounting firm. "Experienced leader with strong integrity and an understanding of organizational and technical needs," it said. "Diligent manager with advanced technical skills capable of rapidly incorporating innovation and change. . . . Technical competence and integrity. . . . National security clearances: Top Secret."

But there was the short run to think about, which Hanssen probably did as he trudged the fifteen or so blocks between the Hoover Building and the State Department, carrying a briefcase full of paperwork. Like most middle-class wage-earners, Bob saw the family's disposable income (and more) disappearing into the new roof, the kids' cars, and general household maintenance. Or maybe, like a lot of middle-aged guys, Hanssen just needed a few more cheap thrills.

He queried the ACS on September 28, October 4, and November 13 in the broadest possible way, asking for all material generated so far that year containing the terms DEAD DROP, WASHINGTON, and DROP SITE.

In mid-November, Hanssen wrote a letter to his SVR contact which was, by turns, arch, flirty, and needy — like a letter to an old lover. In it, Hanssen hinted at the idea of defecting to Russia, collecting his putative "account" and living out his old age training Russian intelligence officers. There was no mention of bringing his wife and children with him. He wrote:

Dear Friends:

On meeting out of the country, it simply is not practical for me. I must answer too many questions from family, friends, and government plus it is a cardinal sign of a spy. You have made it that way because of your policy. Policies are constraints, constraints breed patterns. Patterns are noticed. Meeting in this country is not really that hard to manage, but I am loath to do so not because it is risky but because it

involves revealing my identity. That insulation has been my best pro-
tection against betrayal by someone like me working from whatever
motivation, a Bloch or a Philby. (Bloch was such a shnook. . . . I almost
hated protecting him, but then he was your friend, and there was your
illegal I wanted to protect. If our guy sent to Paris had balls or brains
both would have been dead meat. Fortunately for you he had neither.
He was your good luck of the draw. He was the kind who progressed
by always checking with those above and tying them to his mistakes.
The French said, "Should we take them down?" He went all wet. He'd
never made a decision before, why start then. It was that close. His
kindred spirits promoted him. Things are the same the world over, eh?)

On funds transfers through Switzerland, I agree that Switzerland
itself has no real security, but insulated by laundering on both the in
and out sides, mine ultimately through say a corporation I control
loaning mortgage money to me for which (re)payments are made. . . .
It certainly could be done. Cash is hard to handle here because little
business is ever really done in cash and repeated cash transactions
into the banking system are more dangerous because of the difficulty
in explaining them. That doesn't mean it isn't welcome enough to let
that problem devolve on me. (We should all have such problems, eh?)
How do you propose I get this money put away for me when I retire?
(Come on; I can joke with you about it. I know money is not really put
into an account at MOST Bank, and that you are speaking figuratively
of an accounting notation at best to be made real at some uncertain
future. We do the same. Want me to lecture in your 101 course in my
old age? My college level Russian has sunk low through inattention all
these years; I would be a novelty attraction, but I don't think a practical one except in extremis.)

So good luck. Wish me luck. OK, on all sites detailed to date, but
TOM's signal is unstable. See you in 'July' as you say constant conditions.
yours truly,
Ramon

He seemed to be safe. Bob Hanssen started preparing for his next
drop.

18

ZEROING IN

In the middle of November 2000, someone with a high security clearance arrived at the Hoover Building and was escorted straight to the FBI laboratory on the third floor. A large, strong bag of the kind used by the intelligence agencies to carry classified documents around Washington was received by lab director Donald Kerr and gloved lab personnel, who carried it into a chamber nearly as sterile as an operating theater. National Security Division chief Neil Gallagher and Tim Caruso of the Washington field office had forewarned Kerr that the bag's contents held the solution to an espionage case they had been chasing for nearly a decade. Or so they hoped.

Forensic scientists treated everything in the bag like the evidence from a crime scene. Several large manila envelopes of Russian manufacture were carefully extracted. These were examined closely, photographed, and inventoried. If the material were ever produced for a trial, the record of the people who had touched it — the chain of custody, in legal parlance — had to be precise, so that no one could credibly argue that tampering had occurred. When all the forensic data were compiled, then, and only then, the material was handed over to the team of agents and analysts who waited eagerly in the National Security Division on the fifth floor.

The envelopes were numbered by hand. On a few, there were scribbles in Cyrillic script. The letters had been written by someone who had pulled off one of the most amazing feats in the annals of espionage: he had smuggled a priceless cache of papers and other minutiae out of the fortress-like SVR headquarters building that was known to the U.S. intelligence community as Moscow Center. U.S. officials have carefully guarded the man's identity, but this much is known: he is a former KGB officer recruited by the FBI during the "spies-catch-spies" initiative that had begun with Operation PLAY-ACTOR back in 1991. The ex-KGB man, who may also have worked for the SVR for a spell, was one of many Russian intelligence officers approached by FBI or CIA agents during the mole hunt. The Russian had no idea of the name of the double agent he had known as "B," but he was able to tell the FBI quite a bit about him. He thought "B" had worked at the CIA. The FBI agents on the case and their counterparts at the CIA Counterintelligence Center thought so, too. But that was merely an assumption: the Russian had never actually seen anything that proved the mole was in the CIA.

Sometime in 2000, when the case against the CIA counterintelligence officer who was the prime suspect became stymied for lack of direct evidence, the Russian told his FBI and CIA contacts that the old case files contained some details that might be recognized by those who knew the suspect. The Americans wanted them badly. These files could seal the case against the CIA man or exonerate him. It didn't matter to the Americans which — what they needed was the truth. The Russian was offered a sum of money and permanent sanctuary in the United States, in exchange for which he came up with, or at least agreed to, an audacious plan. He went to Moscow, walked into SVR headquarters, located the storage room containing old KGB case files, and purloined the documents and other items from the case of "B." Then he walked out with them (security at the SVR being, apparently, about as tight as it was at the FBI), and subsequently put the material into envelopes, carefully numbering them to indicate the order in which they should be opened. Into the last envelope he placed a souvenir retrieved from a dead drop on a night years before. Only he knew its significance, so he noted that that envelope had to be kept sealed until he arrived in Washington to explain. He made a rendezvous with CIA officers and handed off the envelopes, to be carried to FBI headquarters, by a method so secure that one official described the exchange as "a terrific story which will never be told."

As the FBI personnel on the case quickly discovered, the material in the envelopes made for a pretty good tale. There were the mole's original letters and the original envelopes in which he had posted them. There was a diary written in Russian by one of the KGB officers who had serviced the dead drops used by the agent "B." There was a tape cassette on which could be heard two voices, a man with a heavy Russian accent and an American. The American said only a few words, and the tape was old and scratchy, but a couple of the analysts thought they recognized the voice.

It was Bob Hanssen.

Bob Hanssen? The FBI's Bob Hanssen? The computer guy? The guy over at State? Yes, the analysts said, they were sure of it. Two of them had worked with Hanssen for years, back when he had been a supervisor in the Soviet Analysis section. They could close their eyes and hear him sitting across the table from them.

And there was more: the letters from "B" and "Ramon Garcia" were laced with phrases that the analysts had heard from Hanssen's lips. One was General Patton's remark about the "purple-pissing Japanese." Hanssen had liked that one, repeated it at work. Then there was the ridiculous advice to KGB officials to study Mayor Daley for tips on how to get and maintain power. That was Bob Hanssen all the way. There were other references to Chicago, Hanssen's hometown. There were even turns of phrase that meant nothing to anybody else but, to analysts who had listened to him every day for years, were pure Hanssenisms.

It was November 16 or 17. Gallagher was in Europe, at a NATO conference, so the mole-hunters advised his deputy, Sheila Horan, of their suspicions of Hanssen. Horan went straight up to "Mahogany

Row," as the Hoover Building's seventh-floor executive suites are called, and alerted Tom Pickard, who was now deputy director of the FBI. Pickard told Freeh that the analysts were still working on it, but they strongly believed the mole was Bob Hanssen. Freeh said little, but his eyes hardened into "the stare," a look that could slice a steel girder. Freeh did not know Hanssen but may have seen him at meetings on intelligence matters and at functions at the Heights School, where one of Freeh's six sons was a student.

On the evening of Friday, November 17, Freeh went to the Heights to give a speech about the importance of balancing family life with a demanding work schedule. He was greeted by an audience of friendly faces, which must have been a welcome respite from the daily pounding of life inside the Beltway. On October 12, the U.S.S. *Cole* had been bombed in the harbor of Aden, and seventeen American sailors had been killed. Freeh had dispatched more than a hundred agents to the scene, under the command of John O'Neill, the New York office's top terrorism expert, but O'Neill had become embroiled in conflicts with Yemeni authorities and U.S. Ambassador Barbara Bodine, neither of whom wanted such a large FBI presence in the volatile region. On a trip to Aden and back in Washington, Freeh spent hours trying to work things out so that O'Neill — a passionate, combative agent known around the office as Take-No-Prisoners-John — could carry out the investigation, which was focusing on a cell of radical Jihadists backed by the terrorist Osama bin Laden. At the same time, Freeh found himself in the uncomfortable position of defending the FBI for its handling of the Wen Ho Lee case, though he knew very well the case had been bungled.

With the November 7 presidential election results still up in the air, the only good news, as far as Freeh was concerned, was that he wouldn't have the Clinton crowd to deal with anymore. He had been persona non grata at the White House ever since he called for an independent counsel to investigate the 1996 Clinton-Gore campaign's fund-raising practices. Clinton aides, and some in Attorney General Janet Reno's circle who resented his agile maneuvering on Capitol Hill, sniped that the FBI director was a self-righteous self-promoter. Of course, if Vice President Gore was declared the victor in 2000, Freeh could not expect much warmer relations with the White House and the Justice Department. Still, his open disdain for Clinton's behavior in the Monica Lewinsky scandal had scored points with the people he really cared about — the FBI's rank and file, and influential conservatives, some of whom were among the families whose sons attended the Heights.

The only sour note on the night of November 17 was that Freeh looked out into the audience and saw Bob Hanssen. Hanssen was sitting there, smiling benignly, as if he were just another loving dad. Freeh was the only one in the room who knew Hanssen's terrible secret. But the FBI director was famous for his stony demeanor. There wasn't a chance Hanssen was going to catch a glimpse of anger in Freeh's coal-black eyes.

Gallagher got the news upon his return to Washington on November 20. Two days later he met with the mole-hunters in a conference room, which had been especially constructed with security features to preclude sound from escaping. "Sit back and listen to what we're going to tell you," one of the investigators said. Gallagher nodded and settled into a chair as the analysts went through twenty-three points of similarity between the verbiage in the letters and Hanssen's idiosyncratic speech patterns. Somber and matter-of-fact, they droned on and on in the flat Joe Friday style to which FBI employees aspire. At some point, they played the tape, which had been determined to date from August 18, 1986. It was the conversation between Hanssen and KGB officer Aleksandr Fefelov from a pay phone in suburban Virginia. Gallagher had walked into the room with questions about the suspected CIA officer. He left absolutely convinced that Hanssen was a double agent. He called the Washington field office and ordered a criminal investigation of Hanssen opened. The next day was Thanksgiving.

True to the sentiments expressed in his speech at the Heights, Freeh left town for a long holiday weekend with his family. Gallagher, Caruso, and the team of agents, analysts, and Gees at the field office did not have the luxury of that choice. They worked shifts on Thanksgiving Day and throughout the weekend. They couldn't afford to let

Hanssen out of their sight for a moment. What if he went to a dead drop and they missed him? Worse, what if he sensed the surveillance and skipped the country? No, they couldn't let that happen. From the moment Gallagher opened the case, the FBI followed Bob Hanssen around the clock. For everybody but the director, whose wrath was storied, family matters had to be put on hold.

On Monday, November 27, Freeh, Pickard, and Gallagher met in the Strategic Information and Operations Center (SIOC, pronouced *sy*-ock), the heavily secured, high-tech crisis command center on the fifth floor of the Hoover Building, not far from the National Security Division. Gallagher walked them through the evidence and laid out a plan of action. The CIA would have to be advised. "I'll call George Tenet," Freeh said. Freeh had become very fond of the CIA director, who was as gregarious and charming as Freeh was stiff and shy.

The subject turned to security. The three men agreed never to mention the case in front of the other assistant directors. In fact, they never even used Hanssen's name in the Hoover Building. Despite their reluctance to use polygraph tests in the past, everybody working on the case would be polygraphed. In time, that number would grow into the hundreds, including not only agents and analysts but also the Gees, laboratory scientists, translators, CIA personnel, and some clerical people.

U.S. Attorney Helen Fahey, based in Alexandria, and John Dion, chief of the Justice Department's Internal Security Section, were briefed. Assistant U.S. Attorney Randy Bellows, working under Fahey, was assigned to handle legal strategy in the case. He applied for and quickly received search and wiretap warrants from the Foreign Intelligence Surveillance Court. Every telephone Hanssen used regularly was bugged. He was put under constant physical surveillance. His work spaces at the State Department and the FBI were searched. His cars were rigged with devices that used GPS technology to enable agents to track his movements remotely.

A financial analysis showed that Hanssen was nowhere as extravagant as Ames had been. Hanssen drove a 1997 Taurus and his wife had a '93 VW van that had to be jump-started regularly by the neighbors. A daughter was on full scholarship to Boston University. The only black mark in his personnel file was that one-week suspension for throwing analyst Kim Lichtenberg to the floor in 1993. Discreet checks around the neighborhood found nothing amiss with his personal life, except for a couple of complaints that he let his dog run around off leash.

The Gees tailed Hanssen's every move. On December 12, they thought they were going to catch him in the act of spying. Hanssen drove past Foxstone Park in Vienna four times. The documents from the KGB indicated that a dead drop named ELLIS was located in the park and had been visited by "B" since 1989. It was only a mile from Hanssen's house on Talisman Drive. All the FCI veterans swore that a pro would never visit a drop site twice. But if Hanssen, who, as an agent, had never spent much time on the streets, showed his inexperience by returning to the same spot, it was hard for the spy watchers to be too cocky about it. He had, after all, successfully eluded the FBI and CIA for two decades. Bob Hanssen may not have been a pro in the minds of the FCI veterans, but he had outsmarted them, and badly. Hanssen didn't stop his car at Foxstone Park; instead, he drove to a shopping center nearby and walked into a store. An SVR officer was standing out front, but the two men didn't speak to each other or show any sign of recognition.

By now, agents investigating the case had gone into the National Security Division database and had discovered audit trails that matched Hanssen's screen name and password. Hanssen, they saw, had been checking the files for quite a while. After so long, and after so much damage, they were finally a step ahead of him. They duplicated the Automated Case Support System in its entirety (for their own use) and entered information about the Hanssen investigation into this new ACS. At the same time, the old database was left in place and monitored by FBI computer experts, who tracked Hanssen using his name and passcode as he searched for clues as to whether he was under suspicion.

All during December, Gallagher and the rest of the mole-hunters had been desperately curious as to what the final, "do-not-open" envelope contained. It remained locked away in the lab, just as the Russian had specified. When the ex-KGB man materialized, just before Christmas, he told his tale at last, and with the appropriate drama. The envelope, he said, contained a shred of a black plastic garbage bag that had been wrapped around some classified documents left at a dead drop years before. He and his colleagues had saved the piece, suspecting that there were fingerprints on it. Since they did not have access to U.S. government employment records, they had nothing with which to compare any prints on the plastic, so they hadn't attempted to lift prints from it. But they banked it, for insurance. Who knew when such a scrap of proof might come in handy?

Well, the day had come. A fingerprint examiner from the lab's latent fingerprints unit lifted two prints from the plastic and compared them to Hanssen's ten-print card, a copy of which had been ordered up from the personnel records section of the FBI's fingerprint identification facility in Clarksburg, West Virginia. The match was positive. Gallagher called Freeh and Pickard. "That iced it," Pickard said.

On the evening of December 26, the Gees saw Hanssen stop three times at the Foxstone Park signal site. On the first pass, he merely paused. The second time, he got out of his car, walked up to the sign with a flashlight, and swept the beam over the sign and nearby pylons. One of the Gees reported that he could be seen "shrugging his shoulders and raising his arms in a gesture of apparent disgust or exasperation." Again, not street-smart. He drove by the signal site again at 9:32 P.M., paused, then headed home.

On January 3, 2001, Hanssen went into the database and ran his name, ROBERT HANSSEN. He found nothing. Six days later, at 8:18 P.M., the Gees watched as Hanssen drove past the Foxstone Park sign. He stopped for ten seconds, apparently looking for a signal, then drove away.

Freeh wanted Hanssen removed from classified access altogether, but this had to be done without alerting him. Freeh, Pickard, and Gallagher cooked up a big "promotion" for Hanssen. They created a fake position working for Gallagher and Bob Dies, head of the Infor-

mation Resources Division on the ninth floor — Siberia. Hanssen
was told that as the most computer-literate counterintelligence
expert in the FBI, he was essential to the upgrading of the National
Security Division database. Moreover, he was being inducted into the
coveted "SES Club." SES stood for Senior Executive Service, meaning,
positions above GS-15. His pay would jump from about $115,000 a
year to about to $131,000. Moreover, he was told, Gallagher wanted
to keep him around past his mandatory retirement date of April 18,
2001. This message was designed to puff Hanssen up in two respects.
Caruso's team had studied Hanssen enough to believe that, like Rick
Ames and Earl Pitts, he had a superiority complex (something the
KGB had figured out years ago). Those who knew Bob thought he
was chronically frustrated by what he perceived as the bureau's failure
to appreciate his true worth. The new job was presented to him as a
belated recognition of his unique and valuable abilities. Second, an
FBI agent's retirement income was figured as a percentage of his
"high threes," his income during his last three years on active duty. A
fat raise and the prospect of a year or two more in the SES Club
would bump Hanssen's pension up by some $15,000 annually. This
was a win-win for Bob Hanssen.

At State, Hanssen's colleagues had no idea he was under surveillance and did what they always did when someone was moving on.
They threw him a farewell party. Fifteen or eighteen people gathered
for lunch at a restaurant of Hanssen's choosing, the *China Garden,*
located in the USA Today Building across the Potomac River in
Arlington, Virginia. His colleagues produced a couple of gag gifts: a
red, white, and blue diplomatic license plate, the vanity kind inscribed
simply BOB. It was mounted on a plaque. There was also a navy base-
ball cap with the letters DSS (for: Diplomatic Security Service).

On January 16, Hanssen again checked into the database,
running the terms DEAD DROP and ESPIONAGE and the dates
12/01/2000–01/15/2001. On January 19, he checked DEAD DROP
12/01/2000–01/18/2001. And on January 22, he ran DEAD DROP
12/01/2000–01/12/2001, DEAD DROP 01/01/2000–01/12/2001, and
FOXSTONE.

He detected no problems. On Tuesday, January 23, the Gees watched him drive past the Foxstone Park signal site, come to a rolling stop nearby, and then drive away. Three days later, shortly after five P.M., he drove slowly past it again.

On January 30, Caruso's team conducted a covert search of Hanssen's Ford Taurus, photographing every item, taking samples, then replacing everything just where it was found. The car was bursting with incriminating evidence. There was a roll of white Johnson & Johnson medical adhesive tape and a box of Crayola colored chalk in the glove compartment, both items mentioned in the "B" documents for use in making signals on signs. In the trunk were seven classified documents printed from an FBI computer, several of which had to do with FBI counterintelligence investigations. There were also six green, fabric-covered, U.S. government ledger notebooks containing classified information, a roll of Superior Performance Scotch clear mailing tape, and some brown-black Hefty garbage bags.

On February 5, agents searched Hanssen's new office, Room 9930 in the Hoover Building. Hanssen's briefcase contained his passport, cell phone, checkbook, and address book. The most interesting item was an eight-megabyte Versa Card Flash Memory Adapter, a memory storage card for a personal computer. Incredibly, it contained several recent letters Hanssen had written to the SVR under his aliases. Was he stupid, arrogant, losing his grip, or hoping to be caught? The agents didn't know or care — they just wanted to wrap up the case, so he wouldn't have a prayer of getting out of jail on a technicality. The agents had their computer specialists download these files, and everything was replaced. That evening, the Gees saw Hanssen drive past Foxstone Park three times between approximately 5:37 P.M. and 7:44 P.M. But once again, he didn't stop.

On Friday night, February 9, Bob and Bonnie joined 550 other parents at Oakcrest for a panel called "Navigating the Perfect Storm of Pop Culture." The Hanssens had come up with the title, and the discussion focused on the challenges of raising wholesome kids in a sex-and-drug-sodden culture. While Oakcrest director Barbara Falk and two fathers spoke, Bob ran the slide projector and made sure the sound worked properly. Afterward, there was a wine and cheese party. "I was surrounded with parents, and Bob went off to get me a glass of Chardonnay," Falk said. "He gave me a tape of the performance, and then we helped clean up. Then when I went to get my briefcase, I found he had put a clean glass of Sprite on it for me to drink going home in the car." Falk was impressed by the gesture.

On Monday, February 12, the Gees found a package wrapped in a piece of garbage bag in the LEWIS drop site in Longbranch Park. They took it to the lab, where it was opened. It contained $50,000 in used one-hundred-dollar bills and a typed note reading: "Next 10/31/01 TOM alt. 20, 27." After being photographed and examined thoroughly, the cash was rewrapped and replaced at the dead drop. That same day, the agents searched Hanssen's Taurus again. This time the glove compartment yielded up a small plastic box containing thumbtacks, which the documents said were used for signaling. There was another signaling item, a piece of pink chalk.

One of the FBI's searches had turned up Hanssen's Palm III organizer. It contained the notations "ELLIS," "8:00," and "February 18." Bingo. Caruso and Fran Burke, head of the SWAT team for Washington field, started working out detailed plans for the takedown. (Burke and at least one more SWAT agent were called in because they had discovered that Hanssen had an impressive collection of firearms, including an AK-47, a shotgun, and a bunch of pistols.) As an added security measure, Caruso decided the actual arrest would be made by two of the youngest, fittest men on his team; the case agents, who were a little older, would wait in the car. The letters the agents had found on the Versa memory card were paranoid and needy — a very scary combination. Caruso and Burke didn't need a psychologist to tell them that Hanssen might be losing his grip. At in-service training sessions at the FBI Academy, agents were regularly lectured on "suicide by cop" scenarios. There were few situations more dangerous for a law enforcement officer than a confrontation with a despondent armed person who did not have the courage to take his own life. Such people often fired at the arresting officer so the officer's partner would be obliged to shoot the shooter. Was Hanssen capable of trying to take out fellow FBI agents? The consensus among Caruso, Burke, and the agents and analysts, who had studied him since November, was: yes. Bob Hanssen was a dark, complicated man. Nobody who knew about the case understood him. Maybe nobody ever would. They had to assume that he was capable of anything.

February 18, 2001, was a Sunday. Freeh, Pickard, and Gallagher decided to meet at the SIOC sometime before eight P.M. to await word of Hanssen's arrest. Gallagher drove to the Hoover Building in the late afternoon, went up to his office, and unwrapped a sandwich. It was going to be a long night, and food and drink were banned from SIOC by techies, who thought some slob of a field agent might dump crumbs or coffee into their computer keyboards.

Gallagher's phone rang at 4:06 P.M. It was the communications center at the Washington field office. "You won't believe it, he's starting to move," the caller said. The next call came at 4:35 P.M. Hanssen had driven to Foxstone Park, parked in the lot, and walked into the woods. The Gees hiding there reported he had reached into his trunk, pulled out one of the Hefty bags, and wrapped a stack of documents, right in front of their eyes.

The hell with SIOC. Gallagher stayed glued to his chair. The comm center's next call came at 4:45 P.M. Hanssen was in handcuffs.

Gallagher thanked the caller, then dialed Freeh's home number. Things had gone down faster than they had expected, he said. Hanssen was already on his way to Tysons Corner.

"Thank you, I'll be right in," Freeh said. That was it. Whatever he was feeling about the treachery of one of the FBI's own he kept to himself. There was an arraignment to prepare for, and there would have to be a press conference. There would be a thank-you ceremony in the FBI auditorium for all the people who had played a part in the investigation. Pickard took small comfort in picturing the uproar at SVR headquarters when the news broke that the "B" files had been nicked. "We're having a bad day today, but Moscow Center's going to have a bad day tomorrow," he told Freeh. Others on the FBI and CIA teams had a damage assessment to do. From what they could tell already, Hanssen could be expected to explain a lot of the unsolved mysteries of the matrix.

As word got around the FBI fraternity, few hesitated to express a sense of profound betrayal. "This is an adult version of school violence," observed Jerry Doyle. "He wasn't bullied. He was just ignored. This was his way of saying, 'Notice me.'"

There was, as well, the terrible realization that the instincts they had honed over a lifetime to smell out lies and evil intentions had failed to pick up the merest inkling about the guy at the next desk. Some wondered if their pride in their own institution, if their belief in the bureau's partly mythologized past had obscured signs of one of the most damaging traitors ever discovered inside the American government.

"I now know how pilots manage to fly their planes into the ground or into the mountain when they're flying on instruments," said Paul Moore, Hanssen's old friend from the analysis unit. "My senses tell me one thing, and the objective input from everything else tells me another."

The spy-hunters had missed the spy.

Viktor Sheymov heard the news on the radio. Now he understood why Hanssen had been so eager to come to work for him. "He was going to steal our source code," he said, referring to the master code underlying the Invicta security system. "The Russians would love to have it." If the Russians got his system, and it worked as well as he thought it did, the NSA might be utterly defeated in its efforts to read the Russian government's communications.

Roy Godson, an influential writer on national security issues, felt the damage extended far deeper and wider than the FBI and even CIA. "I can't think of a case in American history or in the history of espionage in which someone has been at the center of human intelligence gathering and technical intelligence as well as counterintelligence," Godson said. "He taught the Soviets how to beat us, how they could better penetrate and learn our secrets and manipulate them. If you have a human source who's sending you false information, you can check it with a technical source. If you have a compromised technical source, you may be able to find out the truth through a human source. But if your opponent can control both the human and the technical source, he has the ability to decrease in a substantial way your intelligence capability. If you can find out about your opponent's offense and their defense all at the same time, you have the competitive edge."

Even if the Russians themselves never again threatened world peace as they had during the Cold War, Godson said, the secrets Hanssen sold could still do terrible damage to U.S. forces and interests in the Third World. "The Russians are in a position to sell, use, or trade that information to help others neutralize U.S. intelligence," he said. "It's hard to change the way you do things. The Russians know that, and they have taught or traded their knowledge of us to other countries or to non-state actors, to terrorist groups, to revolutionary groups."

It had been almost twenty-two years since Bob Hanssen began selling secrets. But even as the FBI congratulated itself on finally nailing him, there was no denying the presence of an undercurrent of worry, shame, and anger. He had been there. He had been there all along.

19

ENDGAME

Virtually everyone who knew Hanssen was stunned by his arrest. Jack Hoschouer watched the news on Tuesday in his mother's living room in Phoenix. He later read the affidavit and was appalled to find that Bob suggested to the KGB that they try to recruit him — almost immediately after Jack first revealed his distress at being denied promotion. "Within two or three days, he had passed on my name," Jack said, the muscles in his face contorting the way they do when grown men try not to cry. "Paragraph 125 in the affidavit — that's me."

Plato Cacheris understood the stakes immediately. "The government told me on the first day that the death penalty was a very real possibility," Cacheris said. "When this administration says the death penalty, I take it seriously." President George W. Bush, while governor of Texas, had put more prisoners to death than any other governor of any state. It was Attorney General John Ashcroft, a conservative Republican, who first suggested that the government would seek the death penalty for Hanssen. It was necessary, Ashcroft said, "to send a signal that we take very seriously any compromises of the national interest and the national security." Behind the scenes as well, FBI officials said,

Ashcroft stubbornly held out for the death penalty. He was outraged by Hanssen's crimes. It was Freeh, CIA director Tenet, and Secretary of State Donald Rumsfeld who convinced Bush and Ashcroft that the likely disclosure of the nation's lost secrets in a public trial would inflict even worse damage than the renegade FBI agent had already done. Besides, once the government took Hanssen to trial, it would have no power to force him to talk, win or lose. Persuading Hanssen to give a full account was crucial, they argued, because the KGB file, while useful proof of espionage, was not a complete compendium of every secret he had betrayed. There were no details about the classified data Hanssen sold in 1979 and 1980, nor from 1999 until his arrest.

In the end, Cacheris worked out a plea agreement that saved Hanssen from a death sentence and gave the intelligence community what it wanted. But Cacheris could not save Bonnie Hanssen and her children from an uncertain future. The U.S. District Court for the Eastern District of Virginia in Alexandria seized all of Hanssen's property and assets. The court order "enjoined and restrained" family members and associates of Hanssen from hiding or disposing of any assets. That meant, effectively, that Bonnie was prevented from selling anything they owned. No one, least of all she, knew how she would support herself and the children.

Initially, the government argued that Hanssen was in it just for the money. "He betrayed his country, he betrayed his fellow Americans, for no reason other than greed," said Ken Melson, U.S. attorney for the Eastern District of Virginia. But no one really believed that, not even his fellow prosecutors. If it were only cash Hanssen wanted, he would have bargained for much more.

The total that Hanssen had received from Moscow Center for two decades of service came to just about $470,000. The government can account for (at most) $100,000 in jewelry, a used Mercedes, and a trip to Hong Kong that stripper Priscilla Sue Galey made on Hanssen's tab. That leaves about $350,000, which over fifteen years would amount to less than $25,000 a year. Three individuals familiar with different aspects of the case used the same phrase when asked if they had any idea where the money went: "Into the gas tank." The theory, which Cacheris embraced, is that Hanssen probably used the money for daily expenses — gas, groceries, home repair, school and church fees, as well as his children's allowances.

Dr. Alen Salerian, a Washington psychiatrist who had been hired by Cacheris, told the British Broadcasting Corporation on June 17 that Hanssen's espionage had little to do with spying "and much more to do with emotional wounds, all the demons in his own mind that he's been fighting for many, many years." The spying, Salerian said, was an "impulsive escape." Cacheris was furious that the psychiatrist divulged information about Hanssen to the press and fired him. He said that Salerian tried to convince the defense team to mount a psychiatric defense that would discuss Hanssen's interest in sex and pornography; but Cacheris knew that that argument, even if true, would never prevail in front of a jury. "We lost confidence in Salerian," he said. He was especially angry to find out that Salerian had given Hanssen mood-altering prescription drugs. Asked to respond, Salerian said, "I made a recommendation, and the prison doctors, with full knowledge of Plato Cacheris, provided medication."

There is no nice, neat psychiatric diagnosis to explain why Bob Hanssen went so very wrong. One wonders if somewhere along the way, he ever tried to find a respectable way out of his sad and complicated life of lies. There were times when he probably wanted to retrace his steps, undo the damage, pretend, as a child might, that he didn't do it. Maybe he tried to calculate the consequences in 1979, when he spied for the first time, or in 1985, when he resumed for a few years and stopped, or in 1999, when he began spying once more. Maybe he didn't even know when he began the cycle of lies in the first place.

Yet the decision to deal with the devil, to turn over "assets," as the intelligence community so delicately puts it, does not happen on impulse. Perhaps, like an alcoholic or a drug addict, Hanssen wrestled the dragon every day of his adult life, crying out silently for help that never came.

*   *   *

After the arrest, Hanssen spent weeks in solitary confinement at the city jail in Alexandria, Virginia, unable to see a priest. Bonnie and other family members visited him weekly. They would take a bullet-proof elevator to the jail's third floor and walk into the visitor's room, a cramped space lined with cinder blocks, decorated only with a stained maroon rug. There were five partitions, each about two feet wide, where five conversations could take place simultaneously. Visitors sat on small, circular cement seats on one side of a thick, glass window, prisoners on the other. They communicated through a telephone set. Every word was taped. The man who had spent decades keeping a secret now found he had no hope of a private moment.

Over the months, Hanssen grew increasingly despondent. He cried when he talked to visitors. His emotions were raw and very real. He understood that he had betrayed his family, the only people who had ever really loved him. But the answers that everyone sought as to why Hanssen began spying were not easy to find. "People become spies because of an especially intolerable sense of failure as privately defined by themselves," said psychiatrist David Charney before he was hired by the Hanssen defense team. "The pile-up of things reaches intolerable levels. It has to reach a crisis point where the person is desperate, where judgment gets mixed up. The person gets into a major dither and has an epiphany of a marvelous solution that will solve everything. When this happens, he sometimes experiences euphoria. In the next phase, remorse. It always happens, whether a month or a year later. Then there is no way out. You can't turn yourself in because you will never see the kind of teeth and claws as colleagues show people who have gone to the other side."

On Friday morning July 6, 2001, almost five months after the arrest, Hanssen filed into room 800 of U.S. District Court in Alexandria, Virginia. Stooped and gaunt, he had lost at least forty pounds. The green jumpsuit with "PRISONER" stamped on his back hung at his shoulders. His black hair, slightly grayer at the temples, had been poorly cropped by a jailhouse barber.

After taking his seat, Hanssen turned to survey the scene behind him. The courtroom, crowded and noisy, suddenly fell quiet. Several dozen representatives of the FBI's Washington field office sat in the first two rows, their demeanor serious, eyes focused straight ahead. Hanssen's son-in-law, Richard Trimber, an attorney who had helped with the defense, sat at a table in back of the Cacheris team. As Hanssen spotted familiar faces, his reaction was not one of shame. Nor did he show the slightest hint of remorse. Once again, he flashed that totally inappropriate smile, the clenched one seen in the only picture of him that news organizations ever use, the dominant image of one of the worst spies the nation has ever known. "What that smile said was now they have to come to him, to learn from the master how he did it," said the State Department's Ted Strickler. "Now he can teach the experts the tricks of the trade. In his haunted way, he wants his portrait at the top of the heap of turncoats, so they can see his brilliance."

Then Hanssen turned to face the judge and rose to his full height. In that same low voice that forced people to lean forward to hear him, he pleaded guilty to fifteen counts of espionage, attempted espionage, and conspiracy. Under a plea agreement that called for him to spend six months being debriefed about the treachery for which he was paid $1.43 million in cash, diamonds, and foreign bank deposits, he was expected to be sentenced to life in prison with no possibility of parole. Bonnie was to get the survivor's portion of her husband's pension — about $39,000 a year. Legally she was not considered culpable for any knowledge of her husband's treason, and she was allowed to keep the family home and cars.

For the FBI, battered and bruised by scandal after scandal, it was a time of reckoning. Freeh and other senior FBI officials were disgusted, furious that the FBI had been blindsided — and even more embarrassed that they could not assure the nation that there would never again be an FBI turncoat. Freeh asked former FBI director William Webster to head a blue-ribbon panel to delve into the bureau's failure to detect Hanssen's perfidy and control his access to classified material. "If you've lived in the real world, you know there is no absolutely fail-safe setup that will quickly and immediately identify a good man or woman who goes sour," Webster said. "So our focus will be on shortening the distance from detection to defection."

Freeh immediately ordered FBI employees with access to especially sensitive materials to submit to polygraphs and background checks. But he refused to have all agents polygraphed, a recommendation urged upon him as early as 1994 by a succession of top NSD officials, including his former deputy, Bear Bryant, and former Intelligence Division chief John Lewis.

Freeh hesitated because of objections from top-level criminal division officials — and from some within the NSD, as well — that polygraph testing generated an unacceptably high incidence of false positives and inconclusive results, especially among hyper-responsible law enforcement people. Bryant, Lewis, and others argued that while that might be true, the weaknesses of this approach could be remedied with fine tuning; they also pointed to an internal FBI behavioral study in which about forty people convicted of espionage revealed that the thing they all feared most was the polygraph. Perhaps Hanssen had, too, but he'd never had to go on the box.

"It was debated within the bureau for years at the most senior levels," Lewis said. "The downside risks were, this is going to open up Pandora's box, that anybody who has a problem is going to be sidelined, and as we've seen from other organizations, many of these issues you can't resolve, so his career is finished. It was said that some people just can't take a polygraph. I didn't buy that one, but that was a reasonable argument. Also, that it undermined trust and confidence in each other, that the bureau is built on camaraderie, trust, and confidence. I understood that. But you could say the same thing about the CIA or NSA." Both agencies required the periodic polygraphing of their staffs.

Still, Lewis was not confident a polygraph would have caught Hanssen. Professional intelligence agents are trained to beat the polygraph, and people who are convinced of their own rightness —

including aberrant personalities — sometimes don't spike the needle. "You don't catch spies with polygraphs," Lewis acknowledged. "You catch normal people with problems."

Webster, who had always distrusted polygraphs, thought a better deterrent would be a smarter, more compartmentalized computer security system — one that would quietly signal a manager when an employee like Bob Hanssen got into case files he didn't have a need to see. "Invariably [double agents] are apt to wander into areas where they don't belong," Webster said. "We may not always recognize them when they're where they belong, but we can when they aren't.

"In the old days," he added, "we'd have a librarian who'd report it when people asked for files they didn't need to see. We need to have some kind of electronic librarian. Machines can be taught, and I think we can build in a level of uncertainty that makes people in this game hesitate, and that will cut down on their effectiveness."

No machine was capable of seeing through the truly brilliant aspect of Hanssen's tradecraft. By playing the boring-guy-next-door, he completely outfoxed an organization of men and women trained to look for the furtive, the deceitful, the menacing — the other. This would not be the last time the bureau missed a monstrous evil lurking in the banal suburbs. Not one of the nineteen terrorists who hijacked four jetliners and attacked the World Trade Center and the Pentagon on September 11, 2001, was on the FBI's screen — or the CIA's, for that matter — before he crossed the border and disappeared into the American heartland.

The terrorists, like Hanssen, understood that the bureau seldom found those who hid in plain sight. "It amazes me how ordinary these guys looked, yet they ended up being involved in probably the greatest crime in American history," said Corey Moore, assistant manager of Gold's Gym in Greenbelt, Maryland, where five of them got in shape for the hijackings / mass murders.

So it was with Hanssen.

Everyone can draw his own conclusions as to how Hanssen fooled the government for so long. But it is useful to reflect on that moment, less than an hour before his arrest, when Hanssen gave his friend, Jack

Hoschouer, a copy of G. K. Chesterton's book *The Man Who Was Thursday,* the story of seven undercover policemen, each named for a day of the week, who disguised themselves as spies. Hanssen said he thought of himself as the group's poet, Thursday.

The policemen in the novel kept their real identities secret, even from each other, by cloaking themselves in complicated disguises and conducting their meetings in public — lunch on a hotel balcony, dinner in a city restaurant — so no one would ever suspect that they had something to hide. Thursday took all his clues from the president of the organization, a man called Sunday, whom Thursday both admired and feared. Sunday's advice on not being caught: "You want a safe disguise? You want a dress that will guarantee you harmless; a dress in which no one would ever look for a bomb? Why then, dress up as an anarchist. Nobody will ever expect you to do anything dangerous then."

No one ever expected that a man like Robert Philip Hanssen would spy for the Russians. He did not, of course, follow Sunday's advice to the letter and pose as a left-wing activist. No, he was a good family man, a devout Catholic, a member of Opus Dei who attended daily mass and, supposedly, weekly confession, a pious Christian who kept a silver crucifix on his office wall. He was a staunch anticommunist, a right-wing conservative, a proud FBI agent.

It was a safe disguise. Perfect cover.

AUTHORS' NOTE

The Spy Next Door is based on more than 150 interviews with Robert Hanssen's friends, neighbors, colleagues, lawyers, professors, class-mates, roommates, psychiatrists, and priests. Some individuals were interviewed a half-dozen times. Because memories differ, we tried to obtain more than one source for each anecdote. In many instances, we read quotes and passages back to the sources.

When quotation marks were used to recount conversations at which we were not present, at least one person in the conversation repeated the words to us verbatim. In rare instances where a source requested anonymity, the information was used without attribution. But in every case, the person was in a position to know the facts described.

In weighing information, we tried to understand the motives and possible biases of each individual who talked to us. Virtually everyone was perplexed and saddened by the story. Some were angry and felt that the Hanssens, by projecting an image of themselves as model Christians, had deceived them. Most people we contacted simply wanted to make sure we got the record straight on whatever aspect of the case had touched their lives.

Bob and Bonnie Hanssen did not participate in this book. By court order, Bob Hanssen was not permitted to give any news media interviews until his debriefings were complete, which was after this book went to press. Despite dozens of written requests to help us with insight, nuance, and fact checking, Bonnie Hanssen, the Hanssen children, and Janine Brookner, who was Mrs. Hanssen's lawyer, did not grant us interviews.

We have tried to check facts in such manner as we felt reasonable and appropriate. But many details are known to the Hanssen family alone, some locked so deeply in the confines of their hearts that they may not surface. They chose not to confirm or deny any aspect of our reporting. We vowed not to make a surprise visit to Bonnie Hanssen's door or the door of any family member, or to post ourselves at the children's schools. We never did. Other news reporters may disagree with our decision. But in the end, it felt right.

We have proceeded in good faith and take responsibility for any errors.

ACKNOWLEDGMENTS

Every book has a magician hidden in the wings. Ours has been Little, Brown executive editor Geoff Shandler, a wise and creative man who waved his red pencil and pulled a book out of his hat.

We are grateful to our agent, Todd Shuster, of Zachary Shuster Harmsworth, who first approached Elaine with the idea of writing a book about the spy who topped them all.

At *Time*, we would like to thank Managing Editor Jim Kelly and Washington Bureau Chief Michael Duffy, who encouraged and supported this project from the beginning. A special hug to Viveca Novak, Karen Tumulty, and Mark Thompson, who kept us laughing, and to Doug Waller for his insights.

For careful research and endless good cheer, we thank *Time* pros Lissa August and Anne Moffett. A deep bow to our superb copyeditor, Ed Cohen. And immeasurable thanks to Elizabeth Nagle, who helped us turn a manuscript into a book. At Knox College, Bob and Marna Seibert offered historical perspective and were careful readers. Knox College archivist Carley R. Robinson graciously opened the stacks. Retired *Chicago Tribune* reporter John O'Brien tapped his sources at the Chicago Police Department. Sarah Donovan helped with background on Northwestern University Dental School, which graduated its last class in May 2001. Brian Finnerty of Opus Dei did his best to answer our questions. We thank them.

Our biggest thanks go to more than 150 individuals whose lives touched Bob Hanssen: friends, neighbors, classmates, colleagues, and many others, who spent countless hours offering ideas, impressions, history, and anecdotes. Most are named in the book. Several people who provided tremendous guidance and insight asked to remain anonymous. We honor that wish and thank them.

Special thanks from Ann to Henry Putzel, Jr., Mary Lee Jamieson, Ellen Goodman, and Patricia O'Brien, who subjected themselves to unfinished chapters and offered encouragement and direction. And to Margaret Dalton for "walk talk."

Ann also thanks Estrella Damaso, for helping keep her life organized; and always and forever, her husband and children — Mike, Leila, and Christof — whose love, encouragement, and patience know no bounds.

Thanks from Elaine to Dan Morgan and Andrew Shannon Morgan for going it alone at regattas, hikes, kayak outings, bike trips, and hockey games, for patience, and for insisting on smiles. To Astrid Bohm for finding anything, anywhere.

For examples of hope, honor, purpose, and joy, Elaine thanks Edward and Kathryne Shannon, Vesta Mooney, Bill Preston, and Brother Dominic and remembers John Wallace, Phyllis Preston Lee, Robert E. Lee, and Eddie Morgan. We found the place you landed on Omaha Beach. We read about your time on Guadalcanal. We tried to imagine what it was like on those night missions. We thought about how hard you worked, how much you sacrificed, and how long you waited to hear that the war had ended. You are truly the greatest generation, the anchor of our lives, and the beacon.

INDEX